Partially Ordered Algebraic Systems

László Fuchs

Dover Publications, Inc.
Mineola, New York

Bibliographical Note

This Dover edition, first published in 2011, is an unabridged republication of the work originally published by Pergamon Press, Oxford and New York, in 1963.

Library of Congress Cataloging-in-Publication Data

Fuchs, László.
   Partially ordered algebraic systems / László Fuchs. —Dover ed.
     p. cm.
   Originally published: Oxford ; New York : Pergamon Press 1963, in series: International series of monographs on pure and applied mathematics ; 28.
   Includes bibliographical references and index.
   ISBN-13: 978-0-486-48387-0 (pbk.)
   ISBN-10: 0-486-48387-8 (pbk.)
   1. Group theory. 2. Algebraic fields. I. Title.

QA174.2.F83 2011
512'.2—dc23

2011018245

Manufactured in the United States by Courier Corporation
48387801
www.doverpublications.com

TABLE OF CONTENTS

SECOND PART
PARTIALLY ORDERED RINGS AND FIELDS

THIRD PART
PARTIALLY ORDERED SEMIGROUPS

PREFACE

In recent years, interest in the study of partially ordered groups, semigroups, rings and fields has been increasing. The many results in numerous papers are widely spread over the journals, and no systematic survey exists. With the present book I am trying to fill this gap, and to present the most essential results to those who want to become acquainted with this subject.

Algebraic systems endowed with a partial or full order are met with in several disciplines of mathematics. Because the theory of partially ordered algebraic systems is extensive, I have had to abandon the claim to completeness and to be content with developing the main algebraic aspects of the theory. This is the reason why some important topics, such as partially ordered linear or topological spaces, have not been taken into consideration. Moreover, a certain limitation was also necessary in the presentation of results which are purely algebraic in character. To enable the reader to find out more about the subject, references have been provided not only to the original sources of the material here collected, but also to a number of important results which I have not been able to include. No attempt has been made to cover the whole vast field of partially ordered algebraic systems in the bibliography, but for the narrower field surveyed here, the bibliography should be fairly complete.

The text falls into three main parts. I chose the theory of partially ordered groups for the first part, because it is more important both conceptually and from the point of view of the general theory, than the theory of partially ordered semigroups. The second part is devoted to the exposition of partially ordered rings and fields, while the third is concerned with partially ordered semigroups. Some attention has been paid to the non-associative case as well. In order to underline the intrinsic analogies, I endeavoured to set up the three parts on parallel lines as far as possible—this is apparent also from the titles of the chapters.

I have tried to make the presentation self-contained and to give complete proofs of the results. However, in some places it was inevitable to assume some previous knowledge of more or less known results of abstract algebra. In these cases either the needed background is reproduced or references are given. The most familiar concepts of algebra are used, of course, without comment.

My task has been much facilitated by G. BIRKHOFF's standard work *Lattice theory* which I used as a starting point for several chapters. I have also availed myself of the second part of DUBREIL-JACOTIN—LESIEUR—CROISOT's well-known book *Leçons sur la théorie des treillis, des structures algébriques ordonnées et des treillis géométriques.*

I am very grateful to Professor B. H. NEUMANN and Professor L. RÉDEI for their kind help in having read my manuscript and in criticizing it. Prof. NEUMANN's criticisms concerning the language and style of the book were also of great help to me. I wish also to acknowledge the kind help of Professor P. CONRAD in the proof-reading.

I also wish to express my sincere thanks to the Hungarian Academy of Sciences and to its Publishing House for the publication of this volume.

L. F.

Budapest, 31 December 1960

TABLE OF NOTATIONS

$A, B, G, H, R, S, \ldots$	algebraic systems or their subsets
$a, b, c, d, e, u, v, w, x, y, z$	elements of algebraic systems
i, j, k, l, m, n	usually rational integers
ε	1 or -1
$x \in A$	x is an element of the set A
$[x \in A \mid \ldots]$	the set of all $x \in A$ with property $\ldots$
$[x_\lambda]$	the set of the elements x_λ
$A \subseteq B \ (A \subset B)$	A is a (proper) subset of B
$A \cap B, \ A \cup B$	intersection, union of the sets A, B
$A \setminus B$	the set of all x in A but not in B
$\varnothing$	the void set
$a \leq b$	a is less than or equal to b
$a \parallel b$	a and b are incomparable
$a \vee b, \ \vee a_\alpha$	l.u.b. (join) of a and b (of the a_α)
$a \wedge b, \ \wedge a_\alpha$	g.l.b. (meet) of a and b (of the a_α)
$a \perp b$	a and b are orthogonal
a^+, a^-	positive and negative parts of a
$\mid a \mid$	absolute of a
$a : b, \ a :: b$	right and left residuals
$[a, b], \ (a, b)$	closed and open intervals
$U(B), \ L(B)$	sets of upper, lower bounds of B
$P(G), \ G^+$	positive cone of G
$a^\frown$	carrier of a
R_+	additive group of the ring R
$[a, b]$	commutator $a^{-1} b^{-1} ab$ of a and b
$\{S\}$	subgroup generated by S
$\{S\}_\square$	convex subgroup generated by S
$S(a_1, \ldots, a_n), \ S'(a_1, \ldots, a_n)$	normal subsemigroup generated by $a_1, \ldots, a_n$ (and e)
$H(A, a_1, \ldots, a_n)$	semiring generated by $0, A, a_1, \ldots, a_n$
$\cong_o$	order isomorphism
Π, Π^*	direct, complete direct product
Γ	lexicographic product

CHAPTER I

INTRODUCTION

1. Partially ordered sets

It will be convenient to collect here the fundamental concepts and terminology we shall need.

If a binary relation $\leq$ is defined on a set A with the properties

P1. (reflexivity) $\quad a \leq a$
P2. (antisymmetry) $a \leq b$, $b \leq a$ imply $a = b$
P3. (transitivity) $\quad a \leq b$, $b \leq c$ imply $a \leq c$

$\left. \begin{array}{l} \text{for all} \\ a, b, c \in A, \end{array} \right\}$

then A is called a *partially ordered set* (abbreviated: *p. o. set*) and $\leq$ is called a *partial order* on A. The *dual* of A is the p. o. set A' with the same elements and with the partial order $\leq'$ defined as follows: $a \leq' b$ (in A') if, and only if, $b \leq a$ (in A).[1]

As usual, one may write $b \geq a$ for $a \leq b$, and $a < b$ (or $b > a$) to mean that $a \leq b$ and $a \neq b$. If neither $a \leq b$ nor $b \leq a$, then a and b are called *incomparable*, in sign: $a \parallel b$.

It may happen that a relation $\leq$ satisfies only P1 and P3; in this case we say $\leq$ is a *preorder*.[2] A preorder induces an equivalence relation $\sim$ on A, namely, $a \sim b$ if, and only if, simultaneously $a \leq b$ and $b \leq a$. The set of classes a^*, c^*, ... of this equivalence can be partially ordered in the natural way: $a^* \leq c^*$ if for some (and hence for all) a in a^* and for some (and so for all) c in c^* we have $a \leq c$. The set A^* of the classes a^*, c^*, ... is a p. o. set.

A partial order on A induces in the natural way a partial order on any non-void subset B of A; namely, for $a, b \in B$ one puts $a \leq b$ in B if, and only if, $a \leq b$ in the original partial order of A. This *induced* partial order of B will be denoted by the same symbol $\leq$.

A *(closed) interval*[3] $[a, b]$ of A (where $a \leq b$) consists of all $c \in A$ satisfying $a \leq c \leq b$; a and b are called the *endpoints*

---

[1] We shall often speak of the dual of an assertion; thereby we mean that the signs $\leq$ and $\geq$ are to be interchanged throughout.

[2] It is also called *quasiorder;* cf. BIRKHOFF [3].

[3] Generalized intervals have been considered by BURGESS [1].

of $[a, b]$. The subsets $I_a = [x \in A \mid x \geq a]$, defined for each $a \in A$, and their duals $J_a = [x \in A \mid x \leq a]$ are also considered as (closed) intervals. A subset of A is called *convex* if it contains the whole interval $[a, b]$ whenever it contains the endpoints a, b. If we replace $\leq$ by $<$ we obtain the definition of *open intervals* (a, b), ...

Let A and A' be p. o. sets. A mapping $a \to a'$ from A into A' is called *isotone* if it is single-valued and *order-preserving* in the sense that $a \leq b$ implies $a' \leq b'$. A mapping which is one-to-one and isotone in both directions is said to be an *isomorphism* (or *order-isomorphism*) of A onto A'; A and A' are then called *isomorphic* (*order-isomorphic*). If a one-to-one mapping between A and A' reverses order (i. e. $a \leq b$ if, and only if, $a' \geq b'$), then it is a *dual isomorphism*.

Assume that two partial orders, $\leq_1$ and $\leq_2$, are defined on the same set A. Then $\leq_2$ is an *extension* of $\leq_1$ if, for all $a, b \in A$, $a \leq_1 b$ implies $a \leq_2 b$.

A has the *trivial order* if, for all $a, b \in A$, $a \leq b$ implies $a = b$ (i. e. $a < b$ never holds). The order relation $\leq$ is called a *full* (or *linear* or *simple* or *total*) *order* on A and A a *fully ordered* (etc.) *set* (*f. o. set*) or a *chain*, if, in addition to P1—3, also

P4. for all $a, b \in A$, either $a < b$ or $a = b$ or $a > b$

holds. The subsets of a f. o. set are again f. o. sets under the induced partial order.

For a subset B of a p. o. set A, an *upper* (*lower*) *bound* of B in A is an element $u \in A$ ($v \in A$) such that $b \leq u$ ($b \geq v$) for every $b \in B$. B is called *bounded* (in A) if A contains both upper and lower bounds of B. The set of all upper (lower) bounds of B will be denoted by $U(B)$ ($L(B)$). If B consists of the elements $x, y, \ldots$, then we shall often write $U(B) = = U(x, y, \ldots)$ and $L(B) = L(x, y, \ldots)$. If B is the void set, then $U(B) = L(B) = A$, while if B has no upper bound in A, then $U(B) = \varnothing$.

Note that $B \subseteq C$ implies $U(B) \supseteq U(C)$ and $L(B) \supseteq L(C)$. Furthermore

$$L(U(B)) \supseteq B \quad \text{and} \quad U(L(B)) \supseteq B,$$

thus

$$U(L(U(B))) = U(B) \quad \text{and} \quad L(U(L(B))) = L(B).$$

A p. o. set A satisfying

P5. for any $a, b \in A$, the set $U(a, b)$ is not void,

and the dual law

P6. for any $a, b \in A$, the set $L(a, b)$ is not void,

is called u- and l-*directed*, respectively. Obviously, P5 implies that $U(B)$ is never void if B is a finite subset of A. A is said to be *directed* (or to have the *Moore-Smith-property*) if it satisfies both P5 and P6.

A p. o. set A is a $\vee$-*semilattice* or a $\wedge$-*semilattice* according as

P7. to all $a, b \in A$ there exists a $c \in A$ such that $U(a, b) = U(c)$,

or

P8. to all $a, b \in A$ there exists a $d \in A$ such that $L(a, b) = L(d)$

is satisfied. Then the elements

$$c = a \vee b \quad \text{and} \quad d = a \wedge b$$

are uniquely determined elements of A (if they exist) and are called the *join* (or *union* or *least upper bound*) and the *meet* (or *intersection* or *greatest lower bound*)[4] of a and b.

If P7 and P8 are both satisfied, then A is called a *lattice*.[5] A lattice may alternatively be defined as an algebraic system in which two operations, $\vee$ and $\wedge$, are defined such that

L1. $a \vee a = a$ and $a \wedge a = a$,

L2. $a \vee b = b \vee a$ and $a \wedge b = b \wedge a$,

L3. $(a \vee b) \vee c = a \vee (b \vee c)$ and $(a \wedge b) \wedge c = a \wedge (b \wedge c)$,

L4. $(a \vee b) \wedge a = a$ and $(a \wedge b) \vee a = a$

---

[4] We shall use the customary abbreviations l.u.b. and g.l.b.
[5] We shall need several results of lattice theory; the reader is referred to BIRKHOFF [3] for them.

for all $a, b, c \in A$. In fact, the join $a \vee b$ and the meet $a \wedge b$ of a, b satisfy L1—4, and if in some set A the operations $\vee$ and $\wedge$ possess properties L1—4, then putting $a \leq b$ if, and only if, $a \vee b = b$ (or equivalently, $a \wedge b = a$), A becomes a p. o. set in which P7—8 hold.

A f. o. set W is said to be *well-ordered* if every non-void subset V of W contains a smallest element, i. e. a u such that $u \leq v$ for every $v \in V$. The Axiom of Choice will be assumed for all sets. Then by a theorem of ZERMELO, every set can be well-ordered. An equivalent statement is ZORN's lemma: if every subset of a p. o. set A which is a chain (in the induced partial order) has an upper bound in A, then A contains a maximal element, say x, in the sense that if $y \in A$ and $x \leq y$, then $y = x$.

2. Partial order in algebraic systems

As usual, by an *algebraic system* (or *algebraic structure*) we understand a set A in which operations f_a are defined satisfying certain rules.[6] Thus each f_a is a single-valued function from a product set $A \times \ldots \times A$ (say, with $n = n(a)$ components) into A. The notions of isomorphism, homomorphism, etc. of two algebraic systems with the same operations will be understood in the ordinary sense.

A function $g(x)$ from a p. o. set A into a p. o. set A' is called *isotone* if $x \leq y$ in A implies $g(x) \leq g(y)$ in A', and *antitone* if $x \leq y$ in A implies $g(x) \geq g(y)$ in A'. A function $g(x_1, \ldots, x_n)$ of more than one variable can be isotone in some of its variables, antitone in others and both in yet others,[7] and it can, of course, be none of these things in further variables.

The following formulation of monotony is sufficiently general and suitable for all cases to be considered here.

---

[6] We may content ourselves with this naive definition of algebraic systems, since for our purposes we do not need a more precise one.

[7] If A is u-directed or l-directed, then this third alternative means that the function is independent of the variables now considered.

We shall say that an operation f of an algebraic system A satisfies a *monotony law* with the *monotony domain*[8] C if

1. $f(x_1, \ldots, x_n) \in C$ whenever $x_1, \ldots, x_n \in C$,

2. for each i $(i = 1, \ldots, n)$, f is either isotone or antitone or both in the variable $x_i \in C$.

According as which of these three alternatives occurs, f will be called *of type* $\uparrow$, $\downarrow$ or $\updownarrow$ in the variable x_i. We shall say that $f(x_1, \ldots, x_n)$ is of type $(\gamma_1, \ldots, \gamma_n)$ in the domain C if $\gamma_i \, (= \uparrow, \downarrow$ or $\updownarrow)$ denotes the type of f in x_i. If no γ_i equals $\updownarrow$, then f is *non-degenerate* in C.

By a *partially ordered algebraic system* we shall mean a set A satisfying

(i) A is an algebraic system,

(ii) A is a p. o. set,

(iii) every operation f_a of A fulfils a monotony law.

Our main interest lies in groups, rings and semigroups, therefore we shall now derive some consequences of the definition for binary operations.

a) An associative operation $f(x, y)$ for which $g(x, y, z) = f(f(x, y), z)$ (considered as a ternary operation) is non-degenerate must be of type $(\uparrow, \uparrow)$ in any monotony domain C. For assume, on the contrary, that $f(x, y)$ is of type $\downarrow$ in x. Then $g(x, y, z)$ is of type $\uparrow$ in x, while $f(x, f(y, z))$ is obviously of type $\downarrow$ in x. The same argument applies for y.

b) Every operation $f(x, y)$ with a neutral element e (i. e. $f(e, x) = f(x, e) = x$ for all $x \in A$) is of type $(\uparrow, \uparrow)$ in any monotony domain C containing e. In fact, $x_1 < x_2$ implies $f(e, x_1) < f(e, x_2)$ and $f(x_1, e) < f(x_2, e)$.

c) If $f(x, y)$ is of type $(\uparrow, \uparrow)$ and $g(x)$ is a right-inverse operation in the sense that $f(x, g(x))$ is a fixed element e of A for every $x \in A$, then $g(x)$ is of type $\downarrow$ in any C.

Hence we see that in a loop, group, semigroup (with some uninteresting exceptions), and (not necessarily associative) ring with unit element, the multiplication and addition are of type $(\uparrow, \uparrow)$ in every monotony domain; while inversion is of type

---

[8] A monotony domain is always a non-void subset of A.

$\downarrow$ and subtraction is of type ($\uparrow$, $\downarrow$). Thus in the cases mentioned the definition of partial order to be given here is the onl,
reasonable one in view of our general definition of partially
ordered algebraic systems. However, in quasigroups—and *a
fortiori* in groupoids—the operation may be of type ($\uparrow$, $\downarrow$) or
of type ($\downarrow$, $\downarrow$), as shown by the example of integers if the
operation . is defined by $x \cdot y = x - y$ and $x \cdot y = - x - y$,
respectively.

PARTIALLY ORDERED GROUPS

CHAPTER II

PRELIMINARIES ON PARTIALLY ORDERED GROUPS

1. Definitions

A *partially ordered group* (*p. o. group*) is a set G such that

G1. G is a group under multiplication,[1]

G2. G is a p. o. set under a relation $\leq$,

G3. the monotony law[2] holds with the whole of G as monotony domain:

$$a \leq b \text{ implies } ca \leq cb \text{ and } ac \leq bc \text{ for all } c \in G.$$

Since a group contains a neutral element[3] e and the cancellation laws hold, either one of the following conditions is equivalent to G3:

(1) $a \leq b$ implies $cad \leq cbd$ for all $c, d \in G$.

(2) $a < b$ implies $ca < cb$ and $ac < bc$ for all $c \in G$ (that is to say, $x \to cx$ and $x \to xc$ are, for any $c \in G$, one-to-one and isotone mappings).

(3) $a < b$ implies $cad < cbd$ for all $c, d \in G$.

On using the transitivity of $\leq$, it follows readily that further equivalent conditions are the laws:

(4) $a \leq b$ and $a' \leq b'$ imply $aa' \leq bb'$.

(5) $a \leq b$ and $a' < b'$ imply $aa' < bb'$ and $a'a < b'b$.

Let us note the following immediate consequences of the definition:

(6) $a \leq b$ implies $a^{-1} \geq b^{-1}$ [$a < b$ implies $a^{-1} > b^{-1}$].

(7) For all $a, b \in G$, the sets $U(a)$ and $U(b)$ are, regarded as p. o. sets, isomorphic, say, under $x \to ba^{-1}x$.

(8) For all $a \in G$, the sets $U(a)$ and $L(a)$ are, again as p. o. sets, dually isomorphic, say, under $x \to ax^{-1}a$.

---

[1] In most cases it is more convenient to use multiplication rather than addition, in spite of the fact that the formal rules more closely resemble those of addition.

[2] It is often called *homogeneity law*.

[3] e will denote throughout the neutral element of the group.

(9) If G is a p. o. group, then it remains so if the partial order is replaced by its dual.

The following generalizations of the concept of p. o. groups deserve mention.

MATSUSHITA [1] and ZAĬCEVA [2] considered the case when only the half of the monotony law G3 is assumed. Cf. also CONRAD [10] and COHN [1].

A somewhat more general notion than p. o. group has been studied by BRITTON and SHEPPERD [1] under the name "almost ordered groups".

If condition G2 is weakened to the requirement that G is a pre-ordered set, then G3 implies that

$$x \sim y \text{ if, and only if, } xy^{-1} \sim e \text{ (or } y^{-1}x \sim e),$$

and that

$$x_1 \sim y_1, \ x_2 \sim y_2 \text{ imply } x_1 x_2 \sim y_1 y_2.$$

Hence one obtains at once that the equivalence class containing e is a normal subgroup N of G and the other classes are just the cosets of N. The p. o. set of the equivalence classes is nothing else than the factor group G/N which is now a p. o. group.

If $a, b \in G$ have an upper (lower) bound $c \in G$, then their inverses, a^{-1} and b^{-1}, have a lower (upper) bound. Hence a p. o. group which is u-directed (l-directed) is necessarily l-directed (u-directed), and thus directed. If this is the case, we call G a *directed group*.

Moreover, if we assume merely that in the p. o. group G, for some fixed $a_0 \in G$ and for every $b \in G$, an upper (lower) bound of a_0 and b exists, then G is directed. In fact, if a_1 and b_1 are arbitrary and c is an upper (lower) bound for a_0 and $b = a_0 a_1^{-1} b_1$, then $a_1 a_0^{-1} c$ is an upper (lower) bound for a_1 and b_1.

Proposition 1. (CLIFFORD [1].) *If a p. o. group G contains an element $a \geq e$ such that $U(a)$ generates G, then G is directed.*

Conversely, if G is a directed group, then for every $a \in G$ each element b of G may be written in the form[4]

$$b = yz^{-1} \quad \text{with } y, z \in U(a).$$

It suffices to verify the first part for $a = e$, since $U(e) = = a^{-1} U(a)$ is contained in the subgroup generated by $U(a)$. Assuming $\{U(e)\} = G$, any $b \in G$ has the form $b = x_1 \ldots x_r$

---

[4] This is a somewhat sharpened form of CLIFFORD's original result.

with x_i or $x_i^{-1} \in U(e)$. Since the product of two elements of $U(e)$ and the conjugates of every element of $U(e)$ also belong to $U(e)$, b may be written as $b = yz^{-1}$ with $y, z \in U(e)$. Then $y \geq e$ and $y \geq b$, that is, e and each b have an upper bound, and so G is directed. Conversely, if G is a directed group, then $e, b \in G$ have an upper bound $c \in G$. Let $y = ca$ and $z = (b^{-1}c)a$. Then $y, z \in U(a)$ and $b = yz^{-1}$ has the indicated form. Q. E. D.

If $a, b \in G$ have a l. u. b. $a \vee b$ in G, then the inverses a^{-1} and b^{-1} have a g. l. b. $a^{-1} \wedge b^{-1}$ in G. In fact, $(a \vee b)^{-1} \leq a^{-1}$ and b^{-1}, because $a \vee b \geq a, b$, and if $x \leq a^{-1}, b^{-1}$, then $x^{-1} \geq a, b$, whence $x^{-1} \geq a \vee b$, $x \leq (a \vee b)^{-1}$, and $(a \vee b)^{-1}$ is the g. l. b. of a^{-1} and b^{-1}. Hence a p. o. group G which is a $\vee$-semilattice ($\wedge$-semilattice) is at the same time a $\wedge$-semilattice ($\vee$-semilattice) and thus a lattice where

$$a \wedge b = (a^{-1} \vee b^{-1})^{-1} \quad \text{and} \quad a \vee b = (a^{-1} \wedge b^{-1})^{-1}.$$

A p. o. group which is a lattice under its partial order will be called a *lattice-ordered group* (*l. o. group*[5]).

If the order of G is full, we say G is a *fully ordered group* (*f. o. group*).

We list some elementary and useful rules on the sets[6] $U(\ldots, a_a, \ldots)$ and $L(\ldots, a_a, \ldots)$:

(i) $U(\ldots, a_a, \ldots) = \bigcap\limits_a U(a_a)$,

(ii) $xU(\ldots, a_a, \ldots)y = U(\ldots, xa_a y, \ldots)$,

(iii) the multiplication[7] of the $U(\ldots, a_a, \ldots)$ is associative,

(iv) $U(\ldots, a_a, \ldots)^{-1} = L(\ldots, a_a^{-1}, \ldots)$,

(v) $L(a, b) = a\,U(a, b)^{-1}\,b$,

(vi) $U(\ldots, a_a, \ldots)\,U(\ldots, b_\beta, \ldots) \subseteq U(\ldots, a_a b_\beta, \ldots)$,

(vii) $U(x)\,U(\ldots, a_a, \ldots) = U(\ldots, xa_a, \ldots)$,

and the dual laws for all $a_a, b_\beta, x, y \in G$. The proofs are obvious and may be left to the reader.

---

[5] Following BIRKHOFF [1], most authors use the abbreviation *l-group*.

[6] Here the elements a_a may be finite or infinite in number.

[7] The multiplication of the U is defined as multiplication of subsets in the group; similarly for U^{-1}.

If the p. o. group is at the same time an operator group with an operator domain Ω, then it is assumed that the operators $\omega \in \Omega$ preserve order, i. e.

$$a \leq b \text{ implies } a^{\omega} \leq b^{\omega} \text{ for every } \omega \in \Omega.$$

A p. o. group G is said to be *Archimedean* if

$$a^n < b \ (n = 0, \pm 1, \pm 2, \ldots) \text{ implies } a = e,$$

that is, if $\{e\}$ is the only subgroup having an upper bound in G.[8]

A p. o. group is called *completely integrally closed* if

$$a^n < b \ (n = 1, 2, \ldots) \text{ implies } a \leq e.$$

Any completely integrally closed p. o. group is Archimedean. For, if $a^n < b$ ($n = 0, \pm 1, \pm 2, \ldots$), then $a^n < b$ and $(a^{-1})^n < b$ ($n = 1, 2, \ldots$) whence $a \leq e$ and $a^{-1} \leq e$ by complete integral closure; thus $a = e$. The converse implication does not hold in general (see EVERETT and ULAM [1]), but it does in l. o. groups (see Chapter V, 1).

2. The positive cone

In a p. o. group G, an element a is called *positive (integral)* if $a \geq e$, *strictly positive (strictly integral)* if $a > e$, and *negative* if $a \leq e$. If the group operation is written as addition and 0 denotes the neutral element, then positivity has the usual meaning $a \geq 0$.

The set $P = P(G) = G^+$ of positive (integral) elements of G, i. e. $P = U(e)$, is said to be the *positive cone* (or the *integral part*) of G. This concept is a natural tool for studying partial orders. Its precise significance will become fully apparent from Chapter III.

---

[8] Another possibility for defining an Archimedean property is to require that

if $a > e$ and $b > e$ then $a^n > b$ for a suitable $n \geq 1$.

In general, neither of these two Archimedean properties implies the other, but in l. o. groups the one given here implies the one formulated in the text, while in f. o. groups the two definitions are obviously equivalent. — Cf. also JAFFARD [4].

A partial order $\leq$ is already uniquely determined by the corresponding positive cone P, for

(1) $a \leq b$ is equivalent to $ba^{-1} \in P$ (and to $a^{-1}b \in P$).

In view of this, instead of "the partial order with the positive cone P", we may say briefly "the partial order P".

It is readily seen that the reflexivity of $\leq$ is equivalent to $e \in P$, its antisymmetry to the fact that $P \cap P^{-1}$ contains no element $\neq e$, transitivity to the inclusion $PP \subseteq P$, while the monotony law is equivalent to the fact that $ba^{-1} \in P$ implies $(cbd)\,(cad)^{-1} = c(ba^{-1})c^{-1} \in P$, i. e. to $cPc^{-1} \subseteq P$. We may use (1) to define $\leq$ from P if P is given.

Theorem 2. *A subset P of a group G is the positive cone of some partial order of G if, and only if, it satisfies the following three conditions:*

a. $P \cap P^{-1} = e$,
β. $PP \subseteq P$,
γ. $xPx^{-1} \subseteq P$ *for all* $x \in G$.[9]

In other words, P is a normal subsemigroup of G containing e, but no other element along with its inverse.

This result may be completed by

Proposition 3.[10] (a) *G is a directed group if, and only if, P generates G;*

(b) *G is a l. o. group if, and only if, P generates G and P is a lattice under the induced order;*

(c) *G is a f. o. group if, and only if,*

$$P \cup P^{-1} = G.$$

Statement (a) is contained in Proposition 1, while (c) follows at once from the fact that G is f. o. exactly if, for each $a \in G$, either $a \geq e$ or $a^{-1} \geq e$. The "only if" part of (b) is obtained from (a) and from the obvious sublattice property of P. Finally, its "if" part may be shown by verifying that $(ac^{-1} \wedge bc^{-1})c$ is

---

[9] Note that in γ inclusion may evidently be replaced by equality, and because of α, the same is true for β.
[10] Part (a) is due to CLIFFORD [1], (b) to BIRKHOFF [3]. Another form of (c) is: P generates G and is f. o.

a g. l. b. for $a, b \in G$ where c is a lower bound for a, b (note that ac^{-1}, bc^{-1} and $ac^{-1} \wedge bc^{-1} \in P$).

Observe that the partial order on G is trivial if, and only if, P consists of e alone.

BRUCK [1] observes that in a f. o. loop G the sets P and $N = = [x \in G \mid x \leq e]$ have the following characterizing properties: $P \cap N = = e, P \cup N = G; PP \subseteq P; NN \subseteq N; P$ (and hence N) is normal in G.

We turn to an intrinsic characterization of the positive cones. In Theorem 2, P was assumed to be embedded in a group G. We get rid of this assumption in the next

Theorem 4. (BIRKHOFF [1].)[11] *An arbitrary semigroup P is the positive cone of some p. o. group G if, and only if,*

  (i) *the cancellation laws hold in P ;*
  (ii) *P contains a neutral element e ;*
  (iii) *$ab = e$ $(a, b \in P)$ implies $a = b = e$;*
  (iv) *$Pa = aP$ for all $a \in P$.*

We may restrict ourselves to the proof of sufficiency. We embed a semigroup P with properties (i)—(iv) in a group G as follows. If given $a, x \in P$, by (iv) and (i) there exists exactly one $x_a \in P$ such that $xa = ax_a$. The correspondence $x \to x_a$ (with a fixed a) is one-to-one and satisfies

$$A)\ a_a = a, \quad B)\ (xy)_a = x_a\, y_a, \quad C)\ (x_a)_b = x_{ab}.$$

We define G as the set of all pairs (a, b) with $a, b \in P$ subject to the rules:

  a) equality: $(a, b) = (c, d)$ if, and only if, $ad_b = cb$, ·
  b) multiplication: $(a, b) (c, d) = (ac_b, db)$.

Because of A), equality is reflexive. It is symmetric since $ad_b = cb$ implies $ad_b\, d = cbd$. Now by A)—C)

$$d_b\, d = dd_{bd} \quad \text{and} \quad bd = (bd)_{bd} = b_{td}\, d_{td} = b_d\, d_{td},$$

so cancelling by d_{td} we obtain $ad = cb_d$, $(c, d) = (a, b)$. To prove transitivity, let $(a, b) = (c, d)$ and $(a, b) = (g, h)$. Then

[11] For l. o. groups this result is due to J. VON NEUMANN (see BIRK-HOFF [1]).

$ah_b = gb$ implies $ah_b d_b = gbd_b$ where the left member equals $ad_b h_{db} = cbh_{db} = ch_d b$ and the right equals gdb. Cancelling by b we get $ch_d = gd$, $(c, d) = (g, h)$. Equal elements, (a, b) and (c, d), multiplied by (g, h) produce again equal elements. From $iad_b = cb$ we get, on the one hand, that

$$ad_b \, g_{db} \, h_b = cbg_{db} \, h_b, \qquad ag_b \, (hd)_{hb} = cg_d \, hb,$$

i. e. $(a, b)(g, h) = (c, d)(g, h)$, and on the other hand that

$$g(ad_b)_h \, h_{bh} = g(cb)_h \, h_{bh}, \quad ga_h \, (dh)_{bh} = gc_h \, bh,$$

i. e. $(g, h)(a, b) = (g, h)(c, d)$. The associative law of multiplication is a simple consequence of B) and C). It follows at once that (a, a) is a neutral element for multiplication, while (b, a) is inverse to (a, b). Therefore G is a group in which the mapping $a \to (a, e)$ imbeds P isomorphically (and then (a, b) may be interpreted as ab^{-1}). Theorem 2 completes the proof.

Call $a \in G$ *generalized periodic* if there exist conjugates $a_i = x_i^{-1} ax_i$ of a such that $a_1 \ldots a_n = e$. If a belongs to the centre of G, then generalized periodicity is equivalent to being of finite order.

Proposition 5. *The positive cone P does not contain any generalized periodic element $\neq e$.*

If $a \in P$ is generalized periodic, say $a_1 \ldots a_n = e$ where $a_i = x_i^{-1} ax_i \in P$, then e is the product of elements of P. This is possible only if the factors are all equal to e, and it follows that $a = e$.

We shall call the partial order on G *isolated* if $a^n \geq e$ for some natural integer n implies $a \geq e$, and *strongly isolated* if $a_1 \ldots a_n \geq e$ implies $a \geq e$ where again a_i denote conjugates of a. A full order is strongly isolated.

We obtain the following corollaries: A group in which every element is generalized periodic (in particular, a torsion group) admits only the trivial order. A group admitting a (strongly) isolated partial order contains no (generalized) periodic element.

If $\leq_1$ and $\leq_2$ are two partial orders on the same group G and the second is an extension of the first one, then the corresponding positive cones P_1 and P_2 satisfy $P_1 \subseteq P_2$. The converse also holds.

Note that *a necessary and sufficient condition that a partial order P of G be a partial order of a group H containing G is that P remain invariant under all inner automorphisms of H.*

3. Examples

Now we shall illustrate p. o. groups G by a number of examples to some of which we shall refer several times.[12] P will denote throughout the positive cone of G.

1. Let G be the additive group of all integers or rational or real numbers, and let $\leq$ have the usual meaning. G is a f. o. group with Archimedean order.

2. G is the additive group of complex numbers (or complex rational numbers) and let $x + yi$ belong to P if either $x > 0$ or $x = 0$ and $y \geq 0$. Then G is a f. o. group with non-Archimedean order relation.

3. G is the same group, but P is defined differently:

a)[13] $x + yi \in P$ if $x \geq 0$ and $y \geq 0$,

b) $x + yi \in P$ if $x \geq 0$ and $y > 0$, or $x + yi = 0$,

c) $x + yi \in P$ if $x > 0$ and $y > 0$, or $x + yi = 0$,

d) $z = x + yi \in P$, if arc z belongs to the closed (or semi-closed or open) interval $[\alpha, \beta]$ where α and β are fixed angles satisfying $0 \leq \beta - \alpha < \pi$.

All these are Archimedean and directed, the first is l. o.

4. Let G be the multiplicative group of all positive rational numbers and P the set of integers. Then $a \leq b$ means that b/a is an integer, i. e. b is divisible by a. This G is a l. o. group (for in P g. c. d. and l. c. m. exist); it is Archimedean.

5. Let G be the multiplicative group of all non-zero principal ideals (integral and fractional) in a commutative domain of integrity R with unit element, and let P consist of the integral

---

[12] Most of these examples are due to BIRKHOFF [1] and [3].

[13] The additive group of the vectors of the n-dimensional Euclidean space can be f. o. in a similar way to obtain a l. o. group.

principal ideals. G is directed, but in general not a l. o. group. (It is if the unique factorization theorem holds.)

If G is defined as the multiplicative group of the field of quotients of R and P consists of the non-zero elements of R, then we get a pre-ordered group.

6. Define G as the additive group of all real functions f in the interval $[0, 1]$, and let $f \in P$ whenever $f(x) \geq 0$ for all x in $[0, 1]$. Or, take only the continuous functions or functions of bounded variation in $[0, 1]$. In every case we get a l. o. group.

Another example is the same group, but with P consisting of all non-decreasing functions f [i. e. $x \leq y$ implies $f(x) \leq \leq f(y)$] satisfying $f(0) = 0$.

7. G is the additive group of all polynomial functions with real coefficients and domain $[0, 1]$ and P contains those which are non-negative in $[0, 1]$. (A directed but not l. o. group.)

8. Let G be the set of all pairs (x, y) of real numbers under the composition

$$(x_1, y_1) \cdot (x_2, y_2) = (x_1 + x_2, e^{x_2} y_1 + y_2),$$

and let $(x, y) \in P$ mean that either $x > 0$ or $x = 0$ and $y \geq 0$. Then G is a f. o. group, not completely integrally closed.

Another realization of this G is the group of all linear transformations

$$x \to ax + b \quad \text{with real } a, b \text{ and } a > 0$$

under the usual rule of composition (an isomorphism being given by letting $(\log a, b)$ correspond to $x \to ax + b$). A transformation $x \to ax + b$ belongs to P if either $a > 1$ or $a = 1$ and $b \geq 0$.

9. G is the group of real matrices of the form

$$M = \begin{pmatrix} 1 & a & c \\ 0 & 1 & b \\ 0 & 0 & 1 \end{pmatrix}$$

under multiplication, and P consists of the matrices where either $a > 0$ or $a = 0$, $b > 0$ or $a = b = 0$, $c \geq 0$.

10. Let G be the group generated by a, b, c subject to the defining relations

$$ab = ba, \quad ac = cb, \quad bc = ca.$$

Then[14] every element has the unique form $a^n b^m c^k$. We put $a^n b^m c^k \geq e$ if $k > 0$ or $k = 0$ and $n \geq 0$, $m \geq 0$; then G becomes a l. o. group.

4. Subgroups and factor groups

A partial order P on G induces a partial order on a subgroup H of G under which H is obviously again a p. o. group. The positive cone $P(H)$ of this p. o. group H satisfies

$$P(H) = H \cap P(G).$$

This equality could have been taken as the definition of the induced partial order on H.

It is clear that the induced partial order is a full order if the original order is full, but the same need not be true for either directed or lattice-order.[15]

A *convex subgroup* C of G is a subgroup which is a convex subset of G. Some elementary properties are as follows:

(a) A subgroup C is a convex subgroup of G if, and only if, $P(C)$ is a convex subset of $P(G)$.

(b) A convex subgroup of a convex subgroup is convex in the whole group.

(c) The intersection of convex subgroups is again a convex subgroup. Hence we may speak of a convex subgroup generated by a subset X of G; it will be denoted by the symbol $\{X\}_\square$.[16]

(d) If A is a subgroup of G, then

$$\{A\}_\square = AP \cap AP^{-1}$$

---

[14] Note that a, b, c^2 generate an Abelian subgroup of index 2 which is the direct product of the cyclic groups $\{a\}$, $\{b\}$, $\{c^2\}$.

[15] Let G be as in Example 3a and H the subgroup generated by $-1 + i$. Now H is trivially ordered. Note that even if a subgroup of a l. o. group is a lattice with respect to the induced partial order, it need not be a sublattice.

[16] Observe that the convex subgroup generated by a normal subset will again be normal.

in the partial order P of G. To prove this, observe that the obvious equality $AP = PA$ implies that both AP and AP^{-1} are subsemigroups of G and AP^{-1} consists of the inverses of the elements of AP. Thus $AP \cap AP^{-1}$ is a subgroup. If $e \leq$ $\leq x \leq c$ and $c \in AP \cap AP^{-1}$, then $x = ex \in AP$ and $x = c(x^{-1}c)^{-1} \in AP^{-1} \cdot P^{-1} = AP^{-1}$, thus convexity follows. Every element $ax = by^{-1}$ $(a, b \in A; x, y \in P)$ in the intersection belongs to $\{A\}_\square$, since $a \leq ax = by^{-1} \leq b$.

(e) We have $\{X\}_\square = \{X\}$ whenever $P(\{X\}) = e$. In particular, $\{a\}_\square = \{a\}$ provided that $a^n \, || \, e$ for $a^n \neq e$.

(f) Let the subgroup C be directed in its induced partial order. Then C is convex if, and only if, $c \in C$ and $c^{-1} \leq x \leq c$ imply $x \in C$.

A p. o. group containing no normal and convex subgroups other than the trivial subgroups is said to be *o-simple*. The additive group of the real numbers is, for instance, o-simple; there exist even non-commutative f. o. groups which are o-simple.[17]

The following example is due to CLIFFORD [2]. Let G be the group generated by symbols $g(r)$, taken for each rational number r, subject to the defining relations

$$g(r)g(s) = g\left(\frac{1}{2}(r+s)\right)g(r) \quad \text{for } r > s.$$

Then every element $a \neq e$ of G may be written uniquely in the canonical form

$$a = g(r_1)^{m_1} \dots g(r_k)^{m_k} \quad (r_1 < \dots < r_k)$$

with $m_i \neq 0$. Put $a \in P$ if either $a = e$ or $m_k > 0$. Then G is a f. o. group. If H is a normal and convex subgroup of G and $e < a \in H$, then writing a in the canonical form, we have $e < g(r_k) < a^2$ whence $g(r_k) \in H$. If $s < r$ and $g(r) \in H$, then $e < g(s) < g(r)$, $g(s) \in H$. If $s > r$ and $g(r) \in H$, then with $t = 2s - r$ we have

$$g(t)g(r)g(t)^{-1} = g\left(\frac{1}{2}(r+t)\right) = g(s) \in H.$$

Thus $H = G$ and G is o-simple.

[17] This was shown by RIEGER [1] and NEUMANN [1]. CHEHATA [1] has given an example of a f. o. group which is algebraically simple.

It will be useful to call a normal convex subgroup which is directed, an *o-ideal*.

The factor group G/N of a p. o. group G with respect to a normal convex subgroup N can be made into a p. o. group by defining the order relation between the cosets by the rule: $aN \leq bN$ if, and only if, $a' \leq b'$ for some $a' \in aN$, $b' \in bN$.[18] An equivalent definition (showing that our definition is natural) is as follows: $P(G/N)$ is the image of $P(G)$ under the natural homomorphism of G onto G/N. This partial order of G/N may again be called *induced*.[19]

The natural correspondence between the subgroups of G/N and the subgroups of G containing N preserves convexity.

It is worth noticing at this stage:

Proposition 6. (LEVI [3].) *If N is a normal subgroup of an abstract group G, and if both N and G/N are partially ordered, then a necessary and sufficient condition for the existence of a partial order on G inducing those of N and G/N is that $P(N)$ be invariant under all inner automorphisms of G.*

The necessity is obvious. The sufficiency can be proved by verifying directly that the conditions of Theorem 2 are satisfied by the union of $P(N)$ and the cosets of N in G that are in $P(G/N)$ but different from N.[20]

We shall call a p. o. G whose partial order is obtained from those of N and G/N in the manner just described, a *lexicographic extension* of the p. o. group N by the p. o. group G/N.

5. *o*-homomorphisms

Let G and H be p. o. groups. An isotone homomorphism of G into H is called an *o-homomorphism*.[21] The isotony of a homomorphism φ of G into H is equivalent to the fact that φ maps the positive cone $P(G)$ *into* the positive cone $P(H)$:

$$\varphi(P(G)) \subseteq P(H).$$

---

[18] Then to each $a' \in aN$ there exists a $b' \in bN$ such that $a' \leq b'$.

[19] KRULL [2] uses two kinds of partial order in the factor group.

[20] This is not, in general, the only possible definition of $P(G)$. Cf. FUCHS [5].

[21] For a systematic study of *o*-homomorphisms (order-homomorphisms) see ŠIMBIREVA [1] or FUCHS [3].

If, in addition,

$$\varphi(G) = H \text{ and } \varphi(P(G)) = P(H),$$

we call φ an *o-epimorphism* of G (*onto* H).[22] If φ and its inverse φ^{-1} are both *o*-epimorphisms, φ is an *o-isomorphism*, and the groups G and H are *o-isomorphic*, in notation:

$$G \cong_o H.$$

An *o*-isomorphism between G and itself is an *o-automorphism*.

The *o*-automorphisms of a p. o. group G form a subgroup $A_o(G)$ of the automorphism group of G. Also, $A_o(G)$ can be p. o. in case G is directed: we put $a \geq \iota$ for an *o*-automorphism a and the identity map ι of G whenever we have $a^a \geq a$ for all $a \in P(G)$.[23]

Theorem 7. *A normal subgroup N of a p. o. group G is the kernel of an o-homomorphism (o-epimorphism) if, and only if, it is convex. If φ is an o-epimorphism of $\bar{G}$ onto G with kernel N, then*

$$\bar{G} \cong_o G/N.$$

If N is the kernel of an *o*-homomorphism ψ of G, then $a \leq x \leq b$ $(a, b \in N)$ implies $\psi(e) = \psi(a) \leq \psi(x) \leq \psi(b) = \psi(e)$, i. e. $\psi(x) = \psi(e)$, $x \in N$ and N is a convex subgroup. If N is convex, then the natural homomorphism of G onto G/N is, by the definition of the partial order of G/N, an *o*-epimorphism. Finally, if φ is defined as in the theorem, then the correspondence $h \to \varphi^{-1}(h)$ is an *o*-isomorphism between $\bar{G}$ and G/N.

Not both isomorphism theorems can be carried over to p. o. groups. If B and C are subgroups of G and C is normal and convex in $\{B, C\}$, then though $B \cap C$ is normal and convex in B, in general $B/(B \cap C)$ and $\{B, C\}/C$ are not *o*-isomorphic. [Counterexample: G as in Example 3b, $B =$ the real numbers, $C =$ imaginary numbers; then $B/(B \cap C) \cong B$ is

---

[22] This terminology is borrowed from Homological Algebra. It is essential to make distinction between *o*-homomorphism onto and *o*-epimorphism.

[23] For a discussion of $A_o(G)$ see CONRAD [6] and [9]; cf. also COHN [1].

trivially, while $\{B, C\} / C$ is fully ordered.] However, the correspondence $x \to xC$ $(x \in B)$ from $B/(B \cap C)$ onto $\{B, C\}/C$ is readily seen to be isotone.

The analogue of the second isomorphism theorem does hold:[24] *If φ is an o-epimorphism of G onto $\overline{G}$ and $\overline{C}$ is a normal and convex subgroup of $\overline{G}$, then $\varphi^{-1}(\overline{C}) = C$ is normal and convex in G and*

$$G/C \cong_o \overline{G}/\overline{C}.$$

In fact, the composition of φ and the natural mapping from $\overline{G}$ onto $\overline{G}/\overline{C}$ is an *o*-epimorphism, and so the statement follows from Theorem 7.

Note that no analogue of SCHREIER's refinement theorem is true: in the above example the sequences of convex subgroups $G \supset B \supset 0$ and $G \supset C \supset 0$ cannot have a common refinement, since G/B, G/C, C are *o*-simple f. o. groups, while B is trivially ordered.

The analogue of SCHREIER's extension problem in the case of p. o. groups has been discussed by the author [5]. Special cases have been considered by LOONSTRA [5] and by CONRAD [9]. For a discussion of the extension problem for l. o. Abelian groups see JAFFARD [7].

6. Direct products

If the p. o. group G is the direct product of its subgroups A_λ,

$$G = \coprod_{\lambda \in \Lambda} A_\lambda,$$

such that $P(G)$ consists of all elements with λth component in $P(A_\lambda)$ for all λ, then G will be called the *direct* or *cardinal product* of the p. o. subgroups A_λ and $P(G)$ the direct or cardinal product of the $P(A_\lambda)$.

Conversely, if we are given p. o. groups A_λ $(\lambda \in \Lambda)$, then we may construct their direct product and define an element positive when its components are all positive. In the resulting group the original p. o. groups can be *o*-isomorphically embedded (in the natural way) as normal and convex subgroups.

[24]See ŠIMBIREVA [1]. For the f. o. case see RIEGER [1].

Similar definitions apply to the complete direct product II^*A_λ of p. o. groups A_λ, as well as to the $\mathfrak{m}$-direct product[25]

$$C = II^{(\mathfrak{m})}_{\lambda \in \Lambda} A_\lambda$$

which is defined to be the subgroup of the complete direct product consisting of vectors with less than $\mathfrak{m}$ components $\neq e$.

The following statements are obvious.

(a) The $\mathfrak{m}$-direct product of f. o. groups is never f. o. unless at most one component is non-trivial.

(b) The $\mathfrak{m}$-direct product of p. o. groups is l. o. (directed) if, and only if, every factor is l. o. (directed).

(c) The natural map (the projection) of $II^{(\mathfrak{m})} A_\lambda$ onto A_λ is an o-epimorphism.

Theorem 8. (ŠIMBIREVA [1].[26]) *Any two $\mathfrak{m}$-direct decompositions*

$$G = II^{(\mathfrak{m})}_{\lambda \in \Lambda} A_\lambda = II^{(\mathfrak{m})}_{\mu \in M} B_\mu$$

of a directed group G have common refinements. More explicitly, we have

$$A_\lambda = II^{(\mathfrak{m})}_{\mu \in M} C_{\lambda\mu} \quad and \quad B_\mu = II^{(\mathfrak{m})}_{\lambda \in \Lambda} C_{\lambda\mu}$$

where

$$C_{\lambda\mu} = A_\lambda \cap B_\mu.$$

By (b), all of A_λ, B_μ are again directed groups. If P is the positive cone of G, then

$$P_\lambda = A_\lambda \cap P, \quad Q_\mu = B_\mu \cap P$$

are the positive cones of A_λ, B_μ. Define

$$R_{\lambda\mu} = P_\lambda \cap Q_\mu \quad and \quad C_{\lambda\mu} = \{R_{\lambda\mu}\}.$$

---

[25] $\mathfrak{m}$ denotes an infinite cardinal. If $\mathfrak{m} = \aleph_0$ we get the direct product, while if $\mathfrak{m}$ exceeds the power of Λ we obtain the complete direct product.

[26] ŠIMBIREVA proved this theorem only for the direct and the complete direct product. It can be generalized to $\mathfrak{m}$- and $\mathfrak{n}$-direct products as well. — Theorem 8 has been proved for l. o. groups by BIRKHOFF [1].

Evidently $C_{\lambda\mu} \subseteq A_\lambda \cap B_\mu$. If $a \in P_\lambda$, we have $a = a_\mu b_\mu$ with unique

$$a_\mu \in B_\mu, \quad b_\mu \in \prod_{\nu \neq \mu}{}^{(\mathfrak{m})} B_\nu,$$

and $a_\mu = a_{\mu\lambda}\bar{a}_{\mu\lambda}$, $b_\mu = b_{\mu\lambda}\bar{b}_{\mu\lambda}$ with unique

$$a_{\mu\lambda}, b_{\mu\lambda} \in A_\lambda, \quad \bar{a}_{\mu\lambda}, \bar{b}_{\mu\lambda} \in \prod_{\varkappa \neq \lambda}{}^{(\mathfrak{m})} A_\varkappa.$$

Thus $a = (a_{\mu\lambda} b_{\mu\lambda})(\bar{a}_{\mu\lambda} \bar{b}_{\mu\lambda})$ whence — because of the hypothesis $a \in A_\lambda$ — we get $a = a_{\mu\lambda} b_{\mu\lambda}$ and $\bar{a}_{\mu\lambda} \bar{b}_{\mu\lambda} = e$. In view of the definition of partial order in direct products, $\bar{a}_{\mu\lambda}, \bar{b}_{\mu\lambda} \in P$ and therefore $\bar{a}_{\mu\lambda} = \bar{b}_{\mu\lambda} = e$. We infer that $a_\mu = a_{\mu\lambda} \in B_\mu \cap A_\lambda \cap P = R_{\lambda\mu}$, that is,

$$P_\lambda \subseteq \prod_\mu{}^{(\mathfrak{m})} R_{\lambda\mu}.$$

We single out an arbitrary element x of the last $\mathfrak{m}$-direct product and write $x = \langle \ldots, a_\varkappa, \ldots \rangle$ with $a_\varkappa \in P_\varkappa$. Then by convexity

$$a_\varkappa \in \prod_\mu{}^{(\mathfrak{m})} R_{\lambda\mu}$$

for all $\varkappa \in \Lambda$. From $a_\varkappa = \langle \ldots, a_{\varkappa\mu}, \ldots \rangle$ $(a_{\varkappa\mu} \in R_{\lambda\mu})$ we get $a_{\varkappa\mu} \in R_{\lambda\mu} \cap P_\varkappa = e$ and $a_\varkappa = e$ for all $\varkappa \neq \lambda$, thus $x \in P_\lambda$. From

$$P_\lambda = \prod_\mu{}^{(\mathfrak{m})} R_{\lambda\mu}$$

and the fact that the groups are directed we conclude that

$$A_\lambda = \prod_\mu{}^{(\mathfrak{m})} C_{\lambda\mu},$$

and similarly,

$$B_\mu = \prod_\lambda{}^{(\mathfrak{m})} C_{\lambda\mu}.$$

Therefore the inclusion $C_{\lambda\mu} \subseteq A_\lambda \cap B_\mu$ cannot be proper, and this completes the proof.

JAKUBÍK [6] has given a necessary and sufficient condition for arbitrary p. o. groups G that arbitrary two direct decompositions of G with a finite number of components should have a common refinement.

JAKUBÍK [2], [4] proved that in a directed group G every maximal chain C which is convex in G and contains e is a direct factor of G if, and only if, to every $c \in C$ and to every positive $x \in G$, $c \wedge x$ exists in G. This condition is trivially satisfied in l. o. groups.

7. Lexicographic products

There exists another important construction in the theory of p. o. groups which is similar to the direct product and which in the case of f. o. groups is more important.

Let A_λ $(\lambda \in \Lambda)$ be a set of p. o. groups with f. o. index set Λ. Take all elements $a = \langle \ldots, a_\lambda, \ldots \rangle$ of the complete direct product of the A_λ such that the set Λ_a of the indices λ with $a_\lambda \neq e$ is well-ordered (in the order of Λ). We obtain a subgroup of the complete direct product containing the discrete direct product of the A_λ and define $a > e$ whenever $a_{\lambda_0} > e$ for the first element λ_0 of Λ_a. The p. o. group that arises in this way will be called the *lexicographic product*[27]

$$\Gamma A_\lambda \atop \lambda \in \Lambda$$

of the A_λ.

Let σ be an ordinal with the property that the sum and product of any two ordinals smaller than σ are again smaller than σ. In this case the *lexicographic σ-product*[27]

$$G = \Gamma^{(\sigma)} A_\lambda \atop \lambda \in \Lambda$$

of the A_λ may be defined as the subgroup of ΓA_λ consisting of elements a such that the order type of Λ_a is smaller than σ.

In the lexicographic product the components — excepting the last one if it exists — are not convex, but their partial orders are induced by that of the lexicographic product. The subgroups

$$A(\varkappa) = \Gamma^{(\sigma)} A_\lambda \atop \varkappa \leq \lambda$$

plainly are convex in G. $A(\varkappa)$ is the convex subgroup generated by $A_\varkappa$ in G.

Note the following properties.

(a) If G is the lexicographic product of A_1 and A_2, then the projection of G onto A_1 (but not onto A_2) is an *o*-epimorphism.

(b) The lexicographic product of p. o. groups is f. o. exactly if all factors are f. o.

---

[27] Lexicographic products have been used first by HAHN [1] (see Chapter IV, 5). σ-products have been introduced by MAL'CEV [2].

(c) The lexicographic product of non-trivially ordered groups is directed if, and only if, it either has no first factor, or its first factor is directed.

(d) The lexicographic product is l. o. if, and only if, all factors are f. o. except possibly the last one — if such exists — which may be an arbitrary l. o. group.

Theorem 9.[28] *If A_λ and B_μ are directed groups such that*

$$G = \Gamma^{(\sigma)}_{\lambda \leq \Lambda} A_\lambda = \Gamma^{(\sigma)}_{\mu \in M} B_\mu,$$

then the two lexicographic σ-products have o-isomorphic refinements.

We begin the proof with the following

Lemma. *Let*

$$G = \Gamma^{(\sigma)} A_\lambda$$

with directed groups A_λ and let G be the lexicographic product of the directed groups C and D. Then

$$(1) \qquad C \cong_0 \Gamma^{(\sigma)} C_\lambda \quad and \quad D = \Gamma^{(\sigma)} D_\lambda,$$

where C_λ is the component of A_λ in C and $D_\lambda = D \cap A_\lambda$. The groups C_λ and D_λ are again directed.

Assume $g = \langle \dots, a_\lambda, \dots \rangle \in G$ where $a_\lambda \in A_\lambda$, and let λ_0 be the first index λ with $a_\lambda \neq e$. Let $g > e$; then $a_\lambda > e$. Now $g \in D$ if, and only if, $a_{\lambda_0} \in D$, for $g^2 > a_{\lambda_0} > e$ and $a_{\lambda_0}^2 > g > e$. Further $g \in D$ implies $h = \langle \dots, b_\lambda, \dots \rangle \in D$ if the first index λ_1 with $b_{\lambda_1} \neq e$ is greater than λ_0. We infer that $D_\varkappa \neq e$ implies $D_\lambda = A_\lambda$ for $\lambda > \varkappa$ and $D_\varkappa = e$ implies $D_\lambda = e$ for $\lambda < \varkappa$.[29] Thus there may exist at most one D_λ which is a non-trivial subgroup of the corresponding A_λ. Since D is directed, this D_μ must again be directed, thus every D_λ is directed. Our argument shows that $D = \Gamma^{(\sigma)} D_\lambda$.

The o-isomorphism in (1) follows by mapping the elements g of G on their C-components $g(C)$. In fact,

$$g = \langle \dots, a_\lambda, \dots \rangle \to \langle \dots, a_\lambda(C), \dots \rangle$$

[28] This theorem was proved by MAL'CEV [2] for f. o. groups.
[29] The fact that D is directed is essential.

is an o-epimorphism with kernel D which maps G onto $\Gamma^{(\sigma)} C_\lambda$. In order to verify that the C_λ are directed, note that

$$G(\varkappa) = \Gamma^{(\sigma)}_{\varkappa \leq \lambda} A_\lambda \cong_o C_\varkappa^* \times D \quad \text{with} \quad C_\varkappa^* = \Gamma^{(\sigma)}_{\varkappa \leq \lambda} C_\lambda \quad \text{(for } \varkappa \leq \mu),$$

and $G(\ )$ is directed.[30] This completes the proof of the lemma.

Now consider the two σ-products in the theorem, and denote by $C_{\mu\lambda}$ the component of

$$\Gamma^{(\sigma)}_{\nu \geq \mu} B_\nu \cap A_\lambda$$

in B_μ. Apply the lemma to the first σ-product and

$$G = C \times D \quad \text{with} \quad C = \Gamma^{(\sigma)}_{\nu < \mu} B_\nu, \quad D = \Gamma^{(\sigma)}_{\nu \geq \mu} B_\nu$$

to obtain

$$D = \Gamma^{(\sigma)}_\lambda (D \cap A_\lambda)$$

where the groups $D \cap A_\lambda$ are again directed. Applying the lemma again to the last decomposition of D and to

$$D = B_\mu \times \Gamma^{(\sigma)}_{\nu > \mu} B_\nu$$

we get

$$B_\mu \cong_o \Gamma^{(\sigma)}_\lambda C_{\mu\lambda}.$$

Consequently, we obtain the o-isomorphism

$$(2) \qquad G = \Gamma^{(\sigma)}_\mu B_\mu \cong_o \Gamma^{(\sigma)}_\mu \Gamma^{(\sigma)}_\lambda C_{\mu\lambda}.$$

Next we prove that $C_{\mu\lambda} \neq e$ implies $C_{\nu\varkappa} = e$ for $\nu < \mu$ and $\varkappa > \lambda$. Assume, on the contrary, that $C_{\nu\varkappa} \neq e$. Since, by the lemma, the groups $C_{\varrho\tau}$ are directed, there exist $a_\lambda = \langle e, \ldots, e, b_\mu, \ldots \rangle$ in A_λ with $b_\mu > e$ and $a_\varkappa = \langle e, \ldots, e, b_\nu, \ldots \rangle$ in $A_\varkappa$ with $b_\nu > e$. In the lexicographic order, a_λ, $a_\varkappa > e$. Now $\varkappa > \lambda$ implies $a_\varkappa < a_\lambda$, while $\nu < \mu$ implies $a_\varkappa > a_\lambda$, a contradiction. We infer that the two σ-products in the right member of (2) are permutable, i. e.

[30] $C_\lambda = \{e\}$ for $\lambda > \mu$ and $C_\lambda \cong_o A_\lambda$ for $\lambda < \mu$.

$$A_\lambda^* = \Gamma_\mu^{(\sigma)} C_{\mu\lambda}$$

satisfies

$$G \cong_o \Gamma_\lambda^{(\sigma)} A_\lambda^*.$$

Now let $\varkappa$ be any fixed index and $C_{\mu\varkappa} \neq e$ for some μ. For each $e < b_\mu \in C_{\mu\varkappa}$ there exists an

$$a_\varkappa = \langle e, \ldots, e, b_\mu, \ldots \rangle \in \Gamma^{(\sigma)}_{\nu \geq \mu} B_\nu \cap A_\varkappa.$$

By the convexity of

$$A(\varkappa) = \Gamma^{(\sigma)}_{\lambda \geq \varkappa} A_\lambda$$

this group contains every element of $C_{\mu\varkappa}$ between e and b_μ. Since $C_{\mu\varkappa}$ is directed, we conclude that $C_{\mu\varkappa} \subseteq A(\ )$ for every μ, and by convexity, also $C_{\mu\lambda} \subseteq A(\lambda)$ for all $\lambda \geq \varkappa$ and μ. Consequently,

$$A^*(\varkappa) = \Gamma^{(\sigma)}_{\lambda \geq \varkappa} A_\lambda^* = \Gamma^{(\sigma)}_{\lambda \geq \varkappa} \Gamma^{(\sigma)}_\mu C_{\mu\lambda} \subseteq A(\varkappa).$$

Let $a \in A(\varkappa)$. Because of directed order, there exists an $a_\varkappa \in A_\varkappa$ such that $a_\varkappa > a, e$. Now $a_\varkappa = \langle e, \ldots, e, b_\mu, \ldots \rangle$ implies $b_\mu \in C_{\mu\varkappa}$ whence $b_\mu^2 > a_\varkappa > e$ shows that $a_\varkappa \in \{C_{\mu\varkappa}\}_\square \subseteq A^*(\varkappa)$, so that $A^*(\varkappa) = A(\varkappa)$ for every $\varkappa$. We obtain

$$\Gamma^{(\sigma)}_{\lambda > \varkappa} A_\lambda^* = \Gamma^{(\sigma)}_{\lambda > \varkappa} A_\lambda,$$

since these groups are unions of $A^*(\lambda)$ and $A(\lambda)$ with $\lambda > \varkappa$, respectively. Since the order is lexicographic, we arrive at the o-isomorphism $A_\lambda^* \cong_o A_\lambda$. This completes the proof of Theorem 9.

MAL'CEV [2] notes that there exist f. o. groups which cannot be represented as a lexicographic product of f. o. groups having no proper refinements.[31]

8. Intrinsic topologies

There are several ways of defining a topology from the order relation in a p. o. set. If a p. o. group is endowed with

---

[31] The group of Example 8 and the construction in A. KUROSCH, *Gruppentheorie* (Berlin, 1953), pp. 152—153, may be used to give a counterexample.

one or another of the usual topologies, then we do not get a topological group unless some further conditions are imposed. Here we discuss two distinct ways of introducing a topology. The first definition makes the p. o. group into a topological space, but two additional conditions are necessary to ensure the continuity of the multiplication. Though the second immediately implies the continuity of multiplication, it guarantees the closedness of the elements only under a mild hypothesis.

We define the so-called *order topology*.[32] Let G be an arbitrary p. o. group. Assume that $[u_a]$ with u-directed index set is a monotone set[33] in G such that

$$U (. . ., u_a, . . .) = U(a) \quad \text{for some } a \in G.$$

Then we write $u_a \uparrow a$. Dually, we put $v_a \downarrow b$ for a set $[v_a]$ with monotone v_a^{-1} to mean that

$$L (. . ., v_a, . . .) = L(b) \quad \text{for some } b \in G.$$

Then for a set $[x_a]$ with u-directed index set we define

$$x_a \to a$$

(x_a *order converges* or *o-converges* to a, a is the *o-limit* of x_a) if there exist monotone sets u_a and v_a^{-1} such that

$$u_a \leq x_a \leq v_a \quad \text{where } u_a \uparrow a, \; v_a \downarrow a.$$

We have:

(a) *If* $x_a = a$ *for all* a, *then* $x_a \to a$, because we may take $u_a = v_a = a$.

(b) $x_a \to a$ *and* $x_a \to b$ *imply* $a = b$. For if $u_a \leq x_a \leq v_a$, $u_a \uparrow a$, $v_a \downarrow b$, then

$$u_a \in L(. . ., v_a, . . .) = L(b), \quad b \in U(. . ., u_a, . . .) = U(a);$$

hence $b \geq a$. This and the dual argument together lead to $a = b$.

---

[32] BIRKHOFF [3] defines an order topology in the completion of the set by cuts. This method is not adequate in the case of p. o. groups, because p. o. groups cannot in general be embedded in complete p. o. groups.

[33] Monotony means that $\alpha \leq \beta$ implies $u_\alpha \leq u_\beta$.

(c) *If $x_\alpha \to a$ and $[x_\beta]$ is a cofinal[34] subset of $[x_\alpha]$, then $x_\beta \to a$ as well.* On using cofinality, this follows at once from the obvious relations

$$U(\ldots, u_\alpha, \ldots) = U(\ldots, u_\beta, \ldots), L(\ldots, v_\alpha, \ldots) = L(\ldots, v_\beta, \ldots).$$

(d) *If $x_\alpha \to a$ and $y_\alpha \to b$, then $x_\alpha y_\alpha \to ab$.* Let $u_\alpha \leq x_\alpha \leq v_\alpha$, $u'_\alpha \leq y_\alpha \leq v'_\alpha$ where $u_\alpha \uparrow a$, $v_\alpha \downarrow a$, $u'_\alpha \uparrow b$, $v'_\alpha \downarrow b$. Because of **1** (i), (vii) we have[35]

$$U_\alpha(\ldots, u_\alpha u'_\alpha, \ldots) = U_{\alpha,\beta}(\ldots, u_\alpha u'_\beta, \ldots) =$$

$$= \bigcap_\alpha U_\beta(\ldots, u_\alpha u'_\beta, \ldots) = \bigcap_\alpha U(u_\alpha) U_\beta(\ldots, u'_\beta, \ldots) =$$

$$= \bigcap_\alpha U(u_\alpha) U(b) = U(a) U(b) = U(ab),$$

and similarly $L_\alpha(\ldots, v_\alpha v'_\alpha, \ldots) = L(ab)$. Hence $u_\alpha u'_\alpha \leq \leq x_\alpha y_\alpha \leq v_\alpha v'_\alpha$ implies the assertion.

(e) *If $x_\alpha \to a$, then $x_\alpha^{-1} \to a^{-1}$.*

Now call a subset A of G *closed* in the order topology if $x_\alpha \in A$ and $x_\alpha \to a$ imply $a \in A$. Then, G itself and, by (a), the elements of the group are closed, and evidently the intersection of closed sets is again closed. If A and B are closed and $x_\alpha \to c$, $x_\alpha \in A \cup B$, then either the x_α's in A or those in B form a cofinal subset $[x_\beta]$ of $[x_\alpha]$; hence, by (c), c lies either in A or in B, i. e. $A \cup B$ is closed. Hence we have a topological space on G.

In order to guarantee the continuity of multiplication, we have to assume the following two conditions:

(1) If $[x_\alpha]$ is a set in G with u-directed index set such that every cofinal subset contains a cofinal subset o-convergent to a, then x_α also o-converges to a.

(2) Let $[x_{\alpha\beta}]$ be a subset of G, considered as indexed by the pair (α, β) where the index set is u-directed if α or β is kept fixed.[36] If $x_{\alpha\beta} \to a_\alpha$ for every fixed α and $a_\alpha \to a$, then

---

[34] A subset $[x_\beta]$ of $[x_\alpha]$ is cofinal in $[x_\alpha]$ if to each α there exists a β with $\beta \geq \alpha$.

[35] The index of U denotes the varying index.

[36] Here $(\alpha, \beta) \geq (\gamma, \delta)$ means that $\alpha \geq \gamma$ and $\beta \geq \delta$.

there exists a cofinal subset of $[x_{\alpha\beta}]$ which o-converges to a.

Theorem 10. *Let G be a p. o. group in which order convergence satisfies* (1)—(2). *Then G is a topological group under the order topology.*

We already know that, under the order topology, G is a topological space. By (d), aX is closed whenever $a \in G$ and X is a closed subset of G. Since obviously $G \setminus aX = a(G \setminus X)$, we see that aU is open for every open set U.

Now choose a complete system Σ_e of neighbourhoods around e. The elements U_a of Σ_e can be indexed by a u-directed set Λ such that $U_a \subseteq U_\beta$ exactly if $a \geq \beta$ in Λ. Then the open sets aU_a form a complete system of neighbourhoods around a, for if $a \in V$ with some open set V, then $e \in a^{-1}V$ with an open set $a^{-1}V$, thus $U_a \subseteq a^{-1}V$, $aU_a \subseteq V$ for some $U_a \in \Sigma_e$.

Next assume that $[a_a]$ has a u-directed index set and $a_a \to a$. To each open set V containing a there exists an index a_0 such that $a_a \in V$ for $a > a_0$. In fact, $a \notin G \setminus V$ implies that the set of the a_a not in $G \setminus V$ is not a cofinal subset. — Conversely, assume that $[a_a]$ (with a u-directed index set) has the property that to every open set V containing a there exists an index a_0 satisfying $a_a \in V$ for $a > a_0$. If $a_a \to a$ were not true, then because of (1) we could suppose that no cofinal subset of $[a_a]$ o-converges to a. Consider the closure of the set $[a_a]$. There must exist a γ such that the closure X of the set of elements a_a $(a > \gamma)$ does not contain a. In fact, (2) implies that the closure of $[a_a]$ consists of the elements a_a and o-limits of o-convergent subsets of $[a_a]$. If for every γ the subset a_a $(a > \gamma)$ had a subset with o-limit a, then again by (2) a cofinal subset would exist with o-limit a, a contradiction. Now $a \notin X$ implies the existence of an open set V such that $a \in V$ and V, X do not intersect. This contradicts the hypothesis on $[a_a]$, and therefore $a_a \to a$. We conclude that o-convergence and convergence in the sense of order topology are equivalent.

We can now prove the continuity of multiplication. Let $ab^{-1} = c$ and $c \in V$, V an open set. Assume that no pair aU_a, bU_a satisfies $(aU_a)(bU_a)^{-1} \subseteq V$. Then there exist

$a_a \in aU_a$, $b_a \in bU_a$ such that $a_a b_a^{-1} \notin V$. By what has been proved in the preceding paragraph, it follows that $a_a \to a$ and $b_a \to b$; hence, because of (d) and (e), $a_a b_a^{-1} \to ab^{-1}$. We obtain $ab^{-1} \notin V$, a contradiction. This completes the proof of Theorem 10.

Note that in f. o. groups conditions (1) and (2) are always satisfied.

Another way of introducing a topology into a p. o. set A is the interval topology defined by taking the "closed intervals" A, $U(a)$, $L(a)$ (for all $a \in A$) as a sub-base of closed sets. Under this topology, however, the p. o. group fails to be a topological group in very important cases, for instance, in certain l. o. groups. We use a similar method of introducing a topology, one which works under some weak condition, but ensures that the p. o. group will be a topological group. (Its drawback lies in the fact that a l. o. group which is not f. o. will be endowed with the discrete topology.)

Take the subsets

$$G, \ [x \in G \mid x > a], \ [x \in G \mid x < a] \quad \text{for all } a \in G$$

as a sub-base Σ of open sets; i. e., the open sets are the unions of finite intersections of these "open intervals". Σ defines a topology on the space G if, and only if, each point in G is closed. The neutral element e will be a closed subset if, and only if, for every $c \in G$, $c \neq e$, there exists an open set in Σ containing c but not e. If this is not true, then, for some c, $a > c$ always implies $a > e$, and $b < c$ always implies $b < e$; or, otherwise expressed, $P^* c \subseteq P^*$ and $P^* c^{-1} \subseteq P^*$ (where $P^* = P \setminus e$). Hence $P^* c^n = P^*$ for $n = 0, \pm 1, \pm 2, \ldots$. Thus for Σ to define a topology on G, the following condition (*) is necessary:

(*) if $P^* c^n = P^*$ for all integers n, then $c = e$.[37]

---

[37] An equivalent formulation of the condition is: if $a > c^n$ for all $a > e$ and $n = 0, \pm 1, \pm 2, \ldots$, then $c = e$. The additive group of complex numbers with $x + yi > 0$ if, and only if, $x > 0$ does not satisfy(*).

It is readily seen that this is sufficient too. In order to show that G is a topological group under this topology, which may be called the *open-interval topology*, let xy^{-1} belong to an open set of the form (a, b), $a < xy^{-1} < b$. Confining ourselves to the non-discrete case, there exist $c, d \in G$ such that $ay < c < < x < d < by$. Then $V = (c, d)$ is a neighbourhood of x and $W = (b^{-1} d, a^{-1} c)$ is one of y such that $VW^{-1} \subseteq (a, b)$. We obtain:

Theorem 11. *A p. o. group G is a topological group under the open-interval topology if, and only if, G satisfies* (*).

Let us note that (*) is satisfied by l. o. groups. For f. o. groups this is trivially true. In the remaining l. o. groups there exist $a, b > e$ with $a \wedge b = e$,[38] and so $a > c^n$, $b > c^n$ imply $e \geq c^n$ for all n, whence $c = e$.

Proposition 12. *In f. o. groups G the order topology and the open-interval topology are equivalent. They are normal.*

First of all it is evident that in the open-interval topology, the sets $U(a)$ and $L(a)$ form a sub-base for closed sets. Clearly, $U(a)$ and $L(a)$ are closed in the order topology. If X were a non-void subset of G closed in the order topology, but not in the open-interval topology, there would exist an $a \notin X$ such that every interval (b, c) containing a meets X. Hence it is easy to derive a contradiction by standard arguments.

Let A, B be disjoint closed sets in G; then the complement of $A \cup B$ is a union of open intervals (c_a, d_a). Choose $u_a \in (c_a, d_a)$. By hypotheses, every $a \in A$ is separated from B on each side by some (c_a, d_a). For every $a \in A$, adjoin to A the open interval (a, u_a) or (u_a, a) according as which has a meaning. Doing the same with B, we imbed A and B in disjoint open sets. By definition, this is normality.[39]

---

[38] See Chapter V, **4**, Lemma.
[39] For some further results on the topology of f. o. groups we may refer to BAER [2], ISÉKI [2], LOONSTRA [1] and especially to COHEN—GOFFMAN [1].

CHAPTER III

EXTENSIONS OF PARTIAL ORDERS IN GROUPS

1. Extension to a full order

Let $S(a_1, \ldots, a_n)$ denote the normal subsemigroup of a group G that is generated by the elements $a_1, \ldots, a_n (\in G)$, and define $S'(a_1, \ldots, a_n)$ as $S(a_1, \ldots, a_n)$ with e adjoined. These normal subsemigroups will play an important role in dealing with extensions of partial orders P. This is due to the fact that they obey the following obvious rules:

(a) $a \in P$ implies $S'(a) \subseteq P$;

(b) $a \in P$, $a \neq e$, implies $P \cap S(a^{-1}) = \varnothing$;

(c) $S'(a_1, \ldots, a_n) = S'(a_1) \ldots S'(a_n)$;

(d) $S(a_1, \ldots, a_n)^{-1} = S(a_1^{-1}, \ldots, a_n^{-1})$.

The next result has numerous consequences.

Theorem 1. (Fuchs [9].) *A partial order P of a group G can be extended to a full order of G if, and only if, it has the property:*

(*) *for every finite set of elements $a_1, \ldots, a_n$ in G $(a_i \neq e)$ the signs $\varepsilon_1, \ldots, \varepsilon_n$ $(\varepsilon_i = 1$ or $-1)$ can be chosen such that*

$$P \cap S(a_1^{\varepsilon_1}, \ldots, a_n^{\varepsilon_n}) = \varnothing.$$

If P can be extended to a full order Q, then let ε_i be chosen such that $a_i^{-\varepsilon_i} \in Q$. Now $S(a_1^{\varepsilon_1}, \ldots, a_n^{\varepsilon_n})^{-1} = S(a_1^{-\varepsilon_1}, \ldots, a_n^{-\varepsilon_n}) \subseteq Q$, and so

$$P \cap S(a_1^{\varepsilon_1}, \ldots, a_n^{\varepsilon_n}) \subseteq Q \cap S(a_1^{\varepsilon_1}, \ldots, a_n^{\varepsilon_n}) = \varnothing.$$

For the proof of the sufficiency we need

Lemma. *If P satisfies (*) and $a \in G$, then either $PS'(a)$ or $PS'(a^{-1})$ defines a partial order P' in G which again satisfies (*).*

Suppose that G contains elements $a_1, \ldots, a_n, b_1, \ldots, b_m$ $(\neq e)$ such that for every choice of the signs ε_i, η_j one has

$$P \cap S(a, a_1^{\varepsilon_1}, \ldots, a_n^{\varepsilon_n}) \neq \varnothing \quad \text{and} \quad P \cap S(a^{-1}, b_1^{\eta_1}, \ldots, b_m^{\eta_m}) \neq \varnothing.$$

Then the intersection of P with $S(a^\varepsilon, a_1^{\varepsilon_1}, \ldots, a_n^{\varepsilon_n}, b_1^{\eta_1}, \ldots, b_m^{\eta_m})$ is never void, contrary to (*). Thus *either* (i) to every finite set $a_1, \ldots, a_n \ (\neq e)$ in G there are signs $\varepsilon_1, \ldots, \varepsilon_n$ such that

$$P \cap S(a, a_1^{\varepsilon_1}, \ldots, a_n^{\varepsilon_n}) = \varnothing \, ;$$

we then put $P' = PS'(a^{-1})$; *or* (ii) to every finite set $a_1, \ldots, a_n$ $(\neq e)$ in G there are signs $\varepsilon_1, \ldots, \varepsilon_n$ such that

$$P \cap S(a^{-1}, a_1^{\varepsilon_1}, \ldots, a_n^{\varepsilon_n}) = \varnothing \, ;$$

in this case we put $P' = PS'(a)$. (If both (i) and (ii) are true, we can choose either.) Now in case (i), for example, P' is evidently a normal semigroup with e, which moreover satisfies (*); for

$$PS'(a^{-1}) \cap S(a_1^{\varepsilon_1}, \ldots, a_n^{\varepsilon_n}) \neq \varnothing$$

implies

$$P \cap S(a, a_1^{\varepsilon_1}, \ldots, a_n^{\varepsilon_n}) \neq \varnothing \, .$$

Property (*) of P' shows that, for all $b \ (\neq e)$ in G, $P' \cap S(b^\varepsilon) =$ $= \varnothing$ for $\varepsilon = 1$ or -1, i. e. either $b \notin P'$ or $b^{-1} \notin P'$. Thus P' is a partial order of G.

To complete the proof of the theorem, let Q be a maximal element in the set $\mathfrak{P}$ of all partial orders of G which are extensions of P and satisfy (*). Such a Q exists, for (*) is satisfied by the union of an ascending chain of partial orders provided it is satisfied by the members of the chain. By the lemma, for every $a \in G$, either $QS'(a)$ or $QS'(a^{-1})$ also belongs to $\mathfrak{P}$. Therefore $QS'(a)$ or $QS'(a^{-1})$ coincides with Q, i.e. $a \in Q$ or $a^{-1} \in Q$, proving that Q defines a full order. Q. E. D.

Šɪᴋ [5] deals with the case when P has only one linear extension.

2. *O*-groups

Our main concern now is with the groups admitting a linear order. Following Nᴇᴜᴍᴀɴɴ [1] we shall call these groups *O-groups* (orderable groups). A necessary and sufficient condition for having this property can be read directly from Theorem 1.

Theorem 2. (ŁOŚ [1], OHNISHI [2].) *A group G is an O-group if, and only if, given $a_1, \ldots, a_n$ in G with $a_i \neq e$, for at least one choice of the signs $\varepsilon_i = \pm 1$ one has*

$$e \notin S(a_1^{\varepsilon_1}, \ldots, a_n^{\varepsilon_n}).$$

In a group G, the intersection of the 2^n subsemigroups $S(a_1^{\varepsilon_1}, \ldots, a_n^{\varepsilon_n})$ with fixed $a_1, \ldots, a_n$ and varying signs $\varepsilon_1, \ldots, \varepsilon_n$ is either a subgroup or void, therefore another formulation of the last theorem is:

Theorem 3. (LORENZEN [2].) *A necessary and sufficient condition for a group G to be an O-group is that, for every finite set $a_1, \ldots, a_n$ in G ($a_i \neq e$), the intersection of the 2^n semigroups $S(a_1^{\varepsilon_1}, \ldots, a_n^{\varepsilon_n})$ taken for all choices of signs $\varepsilon_i = \pm 1$ is void.*

The property of being an O-group is thus of finite character; consequently,

Corollary 4. (NEUMANN [3].) *In order that G be an O-group it is necessary and sufficient that every finitely generated subgroup of G be an O-group.*

Assume that H is a finitely generated Abelian group. If H is an O-group, then it must be torsion-free. If it is torsion-free, then it is, purely group-theoretically, isomorphic to the lexicographic product of n copies of the f. o. group of the integers, i. e. it is an O-group. Corollary 4 implies

Corollary 5. (LEVI [1].) *An Abelian group is an O-group if, and only if, it is torsion-free.*

Because of the importance of this result, we give an alternative proof for its non-trivial part.

Let G be a torsion-free Abelian group, written additively. By a well-known result,[1] G can be embedded in a torsion-free divisible group D, and it is plainly sufficient to show that D can be f. o. Take a maximal independent set $[g_\alpha]$ in D and order it linearly in some way. Every element $0 \neq x \in D$ may be written uniquely in the form $x = r_1 g_{a_1} + \ldots + r_k g_{a_k}$ with rational coefficients $r_i \neq 0$ and $a_1 < \ldots < a_k$. Define

---

[1] For the results needed from Abelian group theory we refer e. g. to the author's book *Abelian groups* (Budapest, 1958).

x to be positive if $r_k > 0$. It is easy to check that this definition makes D into a f. o. group.

The factor group G/T of an Abelian group G with respect to its maximal torsion subgroup T is torsion-free, hence it admits a full order. This and the trivial order on T produce, by Proposition 6 in Chapter II, a partial order in G which makes G into a directed group unless $T = G$. Thus

Corollary 6. (ŠIMBIREVA [1].) *Every Abelian group containing elements of infinite order admits a directed order.*

A similar argument leads to the conclusion:

Corollary 7. (ŠIMBIREVA [1].) *Every group whose factor-commutator group is not a torsion group admits a directed order.*

From Theorem 1 we get directly:

Corollary 8. *A partial order P of G has no proper extension if, and only if, to each a in G $(a \neq e)$, there exists a finite set of elements, $a_1, \ldots, a_n$ $(\neq e)$ in G such that P intersects $S(a, a_1^{\varepsilon}, \ldots, a_n^{\varepsilon_n})$ for every choice of the signs ε_i.*

Remark. If G is an operator group, then the above results until Corollary 4 remain true provided that $S(a_1, \ldots, a_n)$ denotes the *admissible* normal semigroup generated by the elements $a_1, \ldots, a_n$.

VINOGRADOV [1] proved that free products of O-groups are again O-groups.

NEUMANN and SHEPPERD [1] have shown that if N is a f. o. normal subgroup of finite index of a torsion-free group G and if the inner automorphisms of G leave the positive cone of N invariant, then G can be f. o. so as to extend the ordering of N.

Theorems 2 and 3 have been generalized to right-ordered groups by CONRAD [10].

3. Some group-theoretical properties of O-groups

Because of the importance of O-groups, we pause to consider some group-theoretical properties of O-groups. Here we deal with properties of elements, while those connected with subgroups will be discussed in Chapter IV, **3.**

By Proposition 5 in Chapter II, in an O-group G, $e \in S(a)$ implies $a = e$. Thus an O-group is a generalized torsion-free group.

Proposition 9. (LEVI [2].) *In an O-group G, an equation $x^n = a$ $(\in G)$, n a positive integer, has at most one solution.*

In fact, if x and y were different solutions and, say, $x < y$ in some full order of G, then $x^n < y^n$, a contradiction.

In the next statement, $[a, b] = a^{-1} b^{-1} ab$ means the commutator of a and b, and $[a, b, c] = [[a, b], c]$.

Proposition 10. (NEUMANN [1].) *In an O-group G the following implications hold* :

(i) $[a^m, b^n] = e$ *for some* $m, n \neq 0$ *implies* $[a, b] = e$;

(ii) $[a^m, b, a^n] = e$ *for some* $m, n \neq 0$ *implies* $[a, b, a] = e$.

In view of the identities

$$[a^m, b] = \prod_{i=m-1}^{0} (a^{-i}[a, b]a^i) \qquad \text{for } m > 0$$

and $[a^{-1}, b] = a[a, b]^{-1} a^{-1}$ it follows that $[a^m, b]^\varepsilon \in S([a, b])$ and $[a^m, b^n]^\varepsilon \in S([a^m, b])$ for some $\varepsilon = \pm 1$, thus

$$[a^m, b^n]^\varepsilon \in S([a, b]) \qquad \text{for all } m, n \neq 0.$$

Therefore (i) holds. In the proof of (ii) we use the identity (for $m > 0$):

$$[a^m, b, a] = \Big[\prod_{i=m-1}^{0} a^{-i}[a, b]a^i, a \Big] = \prod_{i=m-1}^{0} t_i^{-1}[a^{-i}[a, b]a^i, a]t_i =$$

$$= \prod_{i=m-1}^{0} (a^i t_i)^{-1} [a, b, a] (a^i t_i)$$

where the t_i are certain products $\Pi a^{-i}[a, b]a^i$. A similar identity is true if $m < 0$; thus $[a^m, b, a^n]^\varepsilon \in S([a^m, b, a]^{\varepsilon'}) \subseteq S([a, b, a])$, whence (ii) follows.

It can be shown that $[a^m, b, c] = e$ $(m > 1)$ is compatible with $[a, b, c] \neq e$ (NEUMANN [1]).

Proposition 11. (LEVI [3].) *Let G be a finitely generated O-group and K the commutator subgroup of G. The factor group G/K contains elements of infinite order.*

Let $a_1, \ldots, a_n$ be the generators and a the greatest among $a_1^{\pm 1}, \ldots, a_n^{\pm 1}$ in some full order of G. Then $\{a\}_\square = G$ and the union of all convex subgroups H not containing a is a proper subgroup of G such that H is normal in G and G/H is o-isomorphic to a real group (cf. Chapter IV, **3**). Hence $K \subseteq H$, and we obtain the assertion.[2]

4. O*-groups

Call a group G an *O*-group* if every partial order of G can be extended to a full order of G.

Theorem 12. (OHNISHI [1].) *A group G is an O*-group if, and only if, G satisfies*

(i) *if $b, c \in S(a)$, then $S(b)$ and $S(c)$ intersect,*
(ii) *$a \neq e$ implies $e \notin S(a)$.*

Let G be an *O*-group. Then (ii) is trivially satisfied. Assume that $b, c \in S(a)$ and $S(b) \cap S(c)$ is void. $P = S'(b) S'(c)^{-1}$ is then the positive cone of a partial order of G. This cannot be extended to a full order Q, for $b \in Q$, $c^{-1} \in Q$, and so $a, a^{-1} \in Q$ would be a contradiction.

Assume that G satisfies (i) and (ii), while a partial order P of G fails to satisfy (*) of Theorem 1, i. e. there exist elements $a_1, \ldots, a_n$ in G ($a_i \neq e$) such that $P \cap S(a_1^{\varepsilon_1}, \ldots, a_n^{\varepsilon_n})$ is never empty. We show that the elements $a_1, \ldots, a_{n-1}$ have again this property. For if not, then for some fixed $\varepsilon_1, \ldots, \varepsilon_{n-1}$ there exist elements $u_1, u_2 \in P$ such that $u_1 = t_1 s_1$, $u_2 = t_2 s_2$ with $t_i \in S'(a_1^{\varepsilon}, \ldots, a_{n-1}^{\varepsilon_{n-1}})$, $s_1 \in S(a_n)$, $s_2 \in S(a_n^{-1})$ where $s_i \neq e$. Because of $s_1, s_2^{-1} \in S(a_n)$, by (i), there exists a b in $S(s_1) \cap S(s_2^{-1})$,

$$b = x_1^{-1} s_1 x_1 \ldots x_k^{-1} s_1 x_k = y_1^{-1} s_2^{-1} y_1 \ldots y_l^{-1} s_2^{-1} y_l.$$

Either $t_1 \neq e$ or $t_2 \neq e$, for otherwise $b \in P$ and $b^{-1} \in P$, whence $b = e$, contrary to (ii). An easy calculation leads us to the conclusion that

$$x_1^{-1} u_1 x_1 \ldots x_k^{-1} u_1 x_k y_l^{-1} u_2 y_l \ldots y_1^{-1} u_2 y_1 \in P \cap S(a_1^{\varepsilon_1}, \ldots, a_{n-1}^{\varepsilon_{n-1}}).$$

Consequently, we may assume $n = 0$. This is impossible.[3]

Corollary 13. (LORENZEN [1], ŠIMBIREVA [1], EVERETT [2].) *An Abelian group is an O*-group if, and only if, it is torsion-free.*

---

[2] For another proof see CHEHATA [3].
[3] Observe that for *O*-groups the analogue of the sufficiency part of Corollary 4 holds true.

In the Abelian case, (ii) is equivalent to being torsion-free, while (i) is always satisfied, for $b = a^m$, $c = a^n$ $(m, n > 0)$, and so $a^{mn} \in S(b) \cap S(c)$.

Note that Theorem 12 holds verbatim if G has an operator domain Ω. The same is true for Corollary 13 provided Ω is assumed to be commutative.

Corollary 14. (PODDERYUGIN [2].) *An Abelian group G with commutative operator domain Ω is an O^*-group if, and only if, G is Ω-torsion-free.*

Naturally we say that G is Ω-torsion-free if $a \neq e$ implies[4] $e \notin S_\Omega(a)$. The proof remains unaltered.

Corollary 15. *A factor group $G/H = G'$ of an O^*-group G is again an O^*-group if, and only if, it satisfies* (ii).

Indeed, (i) is hereditary for factor groups. For if the primes indicate cosets and if $b', c' \in S(a')$, $a \in a'$, then for some $b \in b'$, $c \in c'$ one has $b, c \in S(a)$. If $g \in S(b) \cap S(c)$, then evidently $g' \in S(b') \cap S(c')$.

A further result of some interest may be noted.

Theorem 16. *In an O^*-group G, a normal subgroup C can be written as an intersection of normal subgroups C_λ which are convex subgroups in certain full orders Q_λ of G if, and only if,*

$$(1) \qquad\qquad a \notin C \text{ implies } S(a) \cap C = \varnothing.$$

If C can so be represented and $a \notin C$, then $a \notin C_\lambda$ for some λ. Hence[5] $S(a) \cap C_\lambda = \varnothing$ and the necessity follows. If (1) is satisfied and $a \notin C$,[6] then it follows readily that $P = CS(a) \cup e$ defines a partial order in G. If Q_λ is a full order which is an extension of P, then $CS(a) \cap Q_\lambda^{-1} = \varnothing$. Thus $S(a)$ is disjoint from CQ_λ^{-1} and hence from $C_\lambda = CQ_\lambda \cap CQ_\lambda^{-1}$, the convex subgroup generated by C in Q_λ. This establishes the sufficiency.

MAL'CEV [3] has proved that every locally nilpotent group containing no elements of finite order is an O^*-group. A similar result has been obtained by VINOGRADOV [2].

---

[4] By $S_\Omega(a)$ we denote the admissible normal semigroup generated by a.

[5] Note that G/C_λ admits an induced full order.

[6] $a \notin C$ and $a^{-1} \notin C$ are equivalent.

TEREHOV [1] considered groups G with the property that every full order of every subgroup of G can be extended to a full order of G. He proved that nilpotent groups with this property are necessarily Abelian, and locally solvable groups with this property are solvable of length 2.

5. Intersection of full orders

Theorem 17. (LORENZEN [3].) *A partial order P of a group G is an intersection of full orders if, and only if, $a \notin P$ implies that for every finite set of elements $a_1, \ldots, a_n \in G$ ($a_i \neq e$) there exist appropriate signs $\varepsilon_1, \ldots, \varepsilon_n$ ($\varepsilon_i = \pm 1$) such that*

$$P \cap S(a, a_1^{\varepsilon_1}, \ldots, a_n^{\varepsilon_n}) = \varnothing .$$

The condition is necessary. For if $a \notin P$ and P is the intersection of full orders Q_ν, then $a \notin Q_\nu$ for some ν, i. e. $a^{-1} \notin Q_\nu$ and $PS'(a^{-1})$ can be extended to a full order Q_ν. Thus, by Theorem 1, for a suitable choice of the ε_i we have

$$PS'(a^{-1}) \cap S(a_1^{\varepsilon_1}, \ldots, a_n^{\varepsilon_n}) = \varnothing$$

which implies the stated condition. Conversely, if this condition is fulfilled and $a \notin P$, then $PS'(a^{-1})$ defines a partial order and, by Theorem 1, this has a full extension Q_ν. Now $a^{-1} \in Q$, implies $a \notin Q_\nu$, and thus the intersection of all linear extensions of P equals P.

Corollary 18. *In an O*-group, a partial order P is an intersection of full orders if, and only if, it is strongly isolated, or equivalently,*

$$a \notin P \text{ implies } P \cap S(a) = \varnothing .$$

As in the proof of Theorem 12, it follows that if some a fails to satisfy the condition of the preceding theorem, for some $a_1, \ldots, a_n$, then it does so for $a_1, \ldots, a_{n-1}$. Thus we ultimately reduce n to 0.

Corollary 19.[7] *A necessary and sufficient condition that a partial order P of an Abelian group be the intersection of full orders is that P be isolated.*

This is a simple consequence of Corollary 18.

[7] Cf. e. g. FUCHS [2].

6. Vector groups

A p. o. group G is said to be a *vector group* if it is a subgroup of a (complete) direct product $II^* G_\lambda$ of f. o. groups G_λ, or, otherwise expressed, it is a subdirect product of f. o. groups G_λ. Accordingly, the elements of a vector group are infinite vectors

$$g = \langle \ldots, g_\lambda, \ldots \rangle \quad (g_\lambda \in G_\lambda)$$

such that

$$g \geq e \text{ if, and only if, every } g_\lambda \geq e.$$

The mapping φ_λ: $g \to g_\lambda$ of G onto G_λ is an o-epimorphism.

Lemma. (ŠIMBIREVA [1].) *A p. o. group G is a vector group if, and only if, its positive cone P may be represented in the form*

$$P = \bigcap_\lambda T_\lambda$$

where

1) *the T_λ are normal convex subsemigroups containing P,*
2) $x \in G \setminus T_\lambda$ *implies* $x^{-1} \in T_\lambda$.

If G is a subdirect product of the f. o. groups G_λ, then the $T_\lambda = \varphi_\lambda^{-1}(P_\lambda)$ where P_λ is the positive cone of G_λ satisfy 1)[8] and 2), and the intersection $\cap T_\lambda$ equals P because of the definition of positivity in G. Conversely, if P can so be represented, then the $N_\lambda = T_\lambda \cap T_\lambda^{-1}$ are normal convex subgroups of G, and $\cap N_\lambda = P \cap P^{-1} = e$. Thus G is a subdirect product of the groups $G_\lambda = G/N_\lambda$. If we define the partial order on G_λ by $P_\lambda = \varphi_\lambda(T_\lambda)$, then this is, by 2), a full order, and evidently, $g \in P$ is equivalent to $g \in T_\lambda$ for all λ, i. e. $g_\lambda \in P_\lambda$ for all λ.

Theorem 20. (LORENZEN [3].) *A p. o. group G is a vector group if, and only if, its positive cone P satisfies the following condition:*

$$(1) \qquad \bigcap PS'(a_1^{\varepsilon_1}, \ldots, a_n^{\varepsilon_n}) = P$$

for every finite set $a_1, \ldots, a_n$ of elements of G, where the intersection is to be extended over all possible choices of signs $\varepsilon_i = \pm 1$.

Assume G is a vector group and the T_λ's are chosen as in the lemma. If $a_1, \ldots, a_n \in G$, then to each λ the signs $\varepsilon_1, \ldots,$

---

[8] The convexity follows immediately: $e \leq x \leq a \in T_\lambda$ implies $e = \varphi_\lambda(e) \leq \varphi_\lambda(x) \leq \varphi_\lambda(a)$ in G_λ whence $\varphi_\lambda(x) \in P_\lambda$, $x \in T_\lambda$.

ε_n can be chosen such that we have $a_1^{\varepsilon_1}, \ldots, a_n^{\varepsilon_n} \in T_\lambda$. Hence $PS'(a_1^\varepsilon, \ldots, a_n^{\varepsilon_n}) \subseteq T_\lambda$, and (1) holds. Conversely, if (1) is satisfied by P, and if $a \notin P$, then consider the set of all convex normal semigroups P' containing P such that

$$\cap \, P'S'(a_1^{\varepsilon_1}, \ldots, a_n^{\varepsilon_n})$$

never contains a for any finite set $a_1, \ldots, a_n$ in G. Pick out a maximal T_a in this set. For arbitrary $c \in G$, either $c \in T_a$ or $c^{-1} \in T_a$; for otherwise[9]

$$a \in \cap \, T_a \, S'(c) \, S'(a_1^{\varepsilon_1}, \ldots, a_n^{\varepsilon_n})$$

and

$$a \in \cap \, T_a \, S'(c^{-1}) \, S'(b_1^{\eta}, \ldots, b_m^{\eta_m})$$

for some $\dot{a}_1, \ldots, a_n, b_1, \ldots, b_m$, consequently,

$$a \in \cap \, T_a \, S'(c^{\pm 1}, a_1^\varepsilon, \ldots, a_n^{\varepsilon_n}, b_1^{\eta}, \ldots, b_m^{\eta_m})$$

which is contrary to the choice of T_a. Thus the T_a have the properties 1), 2) of the lemma, and as $a \notin T_a$, the intersection of all T_a is P. An application of the lemma completes the proof.

By virtue of our lemma, a p. o. group whose partial order is an intersection of full orders is necessarily a vector group (where, moreover, each component G_λ may be chosen to be the group itself endowed with a full order). In the case of O^*-groups the converse is also true.

Corollary 21. *For an O^*-group G the property of being a vector group is equivalent to the fact that its partial order is strongly isolated.*

By Corollary 18, it suffices to show that (1) implies that P is strongly isolated. Let $a_1 \ldots a_n = u \in P$ where a_i are conjugates of a. Then $a_1 = u a_n^{-1} \ldots a_2^{-1} \in \cap PS'(a^\varepsilon) = P$ whence $a \in P$. In particular, we obtain

Corollary 22. (CLIFFORD [1].[10]) *A commutative p. o. group is a vector group if, and only if, its partial order is isolated.*

It is easy to show that a l. o. group is, in general, no vector group (ŠIMBIREVA [1]). Let G be the group of Example 10. Here $x = a^{-3} b^5$ satisfies $x = a^2 b^2 c x^{-1} c^{-1}$, thus $x \in PS'(x) \cap PS'(x^{-1})$, but $x \notin P$.

[9] Observe that PS and hence $T_a S$ is convex for every subset S of G.
[10] Cf. also DIEUDONNÉ [1].

CHAPTER IV

FULLY ORDERED GROUPS

1. Archimedean fully ordered groups

By way of introduction we mention an important classification of the elements in f. o. groups.

Let G be a f. o. group. The *absolute* $|a|$ of an element $a \in G$ is defined as $|a| = \max(a, a^{-1})$.

Let $a, b \in G$. The element a is said to be[1] *infinitely small relative to b* if

$$|a|^n < |b| \quad \text{for all positive integers } n; \; .$$

we denote this situation by $a \ll b$. On the other hand, a and b are called *Archimedean equivalent*, in notation: $a \sim b$, if there exist positive integers m and n such that

$$|a| < |b|^m \quad \text{and} \quad |b| < |a|^n.$$

It follows readily that

(i) for each pair $a, b \in G$ one, and only one, of the following relations holds:

$$a \ll b, \qquad a \sim b, \qquad b \ll a;$$

(ii) $a \ll b$ implies $x^{-1} a x \ll x^{-1} b x$ for every $x \in G$;

(iii) $a \ll b$ and $a \sim c$ imply $c \ll b$;

(iv) $a \ll b$ and $b \sim d$ imply $a \ll d$;

(v) $a \ll b$ and $b \ll c$ imply $a \ll c$;

(vi) $a \sim b$ and $b \sim c$ imply $a \sim c$.

Archimedean equivalence separates the elements of G into disjoint classes which we can fully order by defining $\varkappa_1 < \varkappa_2$ for the classes $\varkappa_1$ and $\varkappa_2$ if, and only if, for some (and hence for all) representatives $a \in \varkappa_1$ and $b \in \varkappa_2$ we have $a \ll b$. The element e forms a class by itself, the minimal class; all other classes contain an infinite set of elements, for $a \sim a^n$ ($n =$

---

[1] Another terminology is: a is *incomparably smaller than b*. This concept is often defined only for positive elements.

$= \pm 1, \pm 2, \ldots$). Clearly, G is Archimedean if, and only if, G has not more than two Archimedean classes.[2]

This Archimedean classification can be extended to the positive elements of an arbitrary p. o. group G (LOONSTRA [3]). Let a, b be positive elements of G. Write $a \sim b$ if there exist natural numbers m and n such that $a < b^m$ and $b < a^n$. If there exists an m such that $a < b^m$, but there exists no n such that $b < a^n$, we put $a \lessdot b$. Then the positive cone of the group is again divided into equivalence classes between which a partial order can be defined in a natural manner. See LOONSTRA [3] for some properties of the p. o. set of these Archimedean equivalence classes.

We begin the study of f. o. groups with the Archimedean case. The following result is indispensable in the further development.

Theorem 1. (HÖLDER [1].) *A f. o. group is Archimedean if, and only if, it is o-isomorphic to a subgroup of the additive group of the real numbers with the natural ordering. Thus all Archimedean f. o. groups are commutative.*

The "if" part being obvious, suppose that G is a f. o. group with Archimedean order.

First assume the existence of a $g \in G$ such that $e < g$ and $e \leq x < g$ implies $x = e$. Because of the Archimedean property there exists to each $a \in G$ an integer n such that $g^n \leq a < g^{n+1}$, and so $e \leq a g^{-n} < g$. Hence $a g^{-n} = e$ and $a = g^n$. Consequently, $G = \{g\}$ is o-isomorphic to the group of integers under the mapping $g^n \to n$.

Next assume that no such g exists: to every $x \in G$, $e < x$, we can find a $y \in G$ such that $e < y < x$. Here either $y^2 \leq x$ or $(xy^{-1})^2 \leq x$, thus to each $x > e$ there exists a $z \in G$ with $e < z < x$ and $z^2 \leq x$. Assume that a, b are positive elements of G with $ab \neq ba$, say $ba < ab$. Then put $x = aba^{-1}b^{-1} > e$ and choose a z to this x. By the Archimedean property, there exist natural integers m and n satisfying $z^m \leq a < z^{m+1}$, $z^n \leq b < z^{n+1}$. Hence $x = aba^{-1}b^{-1}$ satisfies $x < z^2$, contrary to $z^2 \leq x$. Thus G is commutative.

In order to construct an o-isomorphism from G into the real numbers, take an $a \in G$, $a > e$, and put $f(a) = 1$. For an

---

[2] Further properties of Archimedean classes may be found in LOONSTRA [2].

arbitrary $b \in G$ consider the sets L, U of all rational numbers m/n $(n > 0)$ such that $a^m \leq b^n$ and $a^m > b^n$, respectively. By the Archimedean property neither L nor U is void, and it is readily seen that (L, U) defines a Dedekind section of the rational numbers. If the real number β corresponds to this section, then write $f(b) = \beta$. We have $f(bc) = f(b) + f(c)$; for if m/n and r/s $(n, s > 0)$ belong to the lower classes defined by b and c, respectively, then $a^m \leq b^n$, $a^r \leq c^s$. This implies $a^{ms+nr} \leq$ $\leq (bc)^{ns}$, i. e. $(m/n) + (r/s)$ belongs to the lower class defined by bc. A similar argument applies to the upper classes. Thus f is a homomorphism of G into the real numbers. Since $f(b) = 0$ is inconsistent with $b > e$ (then $a \leq b^n$ for some n and so $1/n$ would lie in the lower class of b), and since f is obviously isotone, we conclude that f is an o-isomorphism from G into the real numbers. Q. E. D.

This proof is based on CARTAN's proof [1]; cf. also BAER [2], RIEGER [1], SCHILLING [1]. Another proof is given by LEVI [2]. (Cf. Ch. XI where a corresponding result on semigroups is proved.) CHEHATA [3] shows that in a f. o. group the commutator of two elements is infinitely small relative to the greater of the two whence the commutativity of Archimedean f. o. groups follows at once. (Cf. also the proof of Theorem 18 in Chapter V.)

For generalizations to loops see ISÉKI [1] and PICKERT [2]. Cf. also ZELINSKY [1].

The problem of deciding the o-isomorphy of two subgroups of the real numbers now naturally presents itself. A complete answer may easily be given:

Proposition 2. (HION [1].) *Let $A \neq 0$ and B be sub-groups of the additive group of real numbers, endowed with the natural ordering, and φ an o-homomorphism (or o-isomorphism) from A into B. Then there exists a real number $r \geq 0$ such that*

$$\varphi(a) = ra \quad \text{for all } a \in A.$$

If[3] $\varphi(a_0) = 0$ for some $a_0 \in A$, $a_0 > 0$, then $\varphi(a) = 0$ for all $a \in A$, because $0 < a < na_0$ implies $\varphi(a) = 0$. In this case $r = 0$. In the remaining case $a_i > 0$ $(a_i \in A)$ implies $\varphi(a_i) > 0$. Assume e. g. that $\varphi(a_1) : \varphi(a_2) < a_1 : a_2$, and take a rational

---

[3] We use additive notation in A and B.

number m/n $(m, n > 0)$ between $\varphi(a_1)\,/\,\varphi(a_2)$ and a_1/a_2. We have $ma_2 < na_1$, while $m\,\varphi(a_2) > n\,\varphi(a_1)$, which is impossible. Thus $\varphi(a) : a = r$ is a constant and is plainly > 0.

Proposition 2 holds for subsemigroups of the real numbers as well.

Corollary 3. *The o-automorphisms of an Archimedean f. o. group form a subgroup of the multiplicative group of positive real numbers.*

The group $A_0(G)$ of all o-automorphisms of f. o. Abelian groups G has been discussed by CONRAD [6]. He established certain sufficient conditions under which $A_0(G)$ can be fully ordered. Cf. also COHN [1], CONRAD [9].

Finally, we mention two easy consequences of the main theorem.

Corollary 4. (LOONSTRA [1].) *A continuous f. o. group $G \neq e$ (i. e. every Dedekind section determines one, and only one, element) is o-isomorphic to the f. o. group of the real numbers.*

Assume that $a^n \leq b$ for some $a \geq e$ and for $n = 0, 1, 2, \ldots$. The assumption implies that $\lim \sup a^n = x$ exists, that is to say, $U(e, a, \ldots, a^n, \ldots) = U(x)$. By Chapter II, **1** (ii), this x satisfies $ax = x$ whence $a = e$ and the group has Archimedean order. By Theorem 1 it is o-isomorphic to a subgroup of the real numbers, and this must again be continuous.

Corollary 5. (FAN [1], FUCHS [3], MICHIURA [8].) *If the partial order of G is isolated and G has no convex subgroup other than G and e, then G is o-isomorphic to a subgroup of the real numbers (and hence fully ordered).*

Let $a \in G$, $a \parallel e$. Then, by the assumption of isolatedness, a is of infinite order and again $a^n \parallel e$. Thus $\{a^2\}$ is a nontrivial convex subgroup. Therefore G is f. o. If $a^n < b$ for $n = 0, \pm 1, \pm 2, \ldots$, then b does not belong to $\{a\}_\square$, thus $a = e$ and the order is Archimedean. A simple appeal to Theorem 1 completes the proof.

2. Full orders on free groups

The main purpose of this section is to prove that every f. o. group G is an o-epimorphic image of a f. o. free group. The key idea in this context (to be used extensively in the next section) is that of establishing a certain chain of subgroups with the property that the factor groups defined by successive members are easily orderable.

A descending chain

$$G = C_0 \supset C_1 \supset \ldots \supset C_a \supset C_{a+1} \supset \ldots$$

with a varying over ordinals less than a fixed τ is called a *transfinite central series* of G if C_{a+1} is a (normal) subgroup of C_a such that the commutator $[G, C_a]$ is contained in C_{a+1} and, for a limit ordinal a, C_a is the intersection of all C_β with $\beta < a$. Clearly, the C_a are normal in G.

Theorem 6. (ŠIMBIREVA [1], NEUMANN [1].) *If a group G has a transfinite central series ending with $C_\tau = e$ such that all factor groups C_a/C_{a+1} are torsion-free, then G is an O-group.*

The definition implies that the C_a/C_{a+1} are Abelian, thus by LEVI's theorem (Chapter III, Corollary 5) there exist full orders P_a on C_a/C_{a+1}. Define the set P in G to consist of e and of all $g \in G$ ($g \neq e$) with the property: if a is the ordinal defined by[4] $g \in C_a \setminus C_{a+1}$, then the coset gC_{a+1} belongs to P_a. Then for a $g \in G$ either $g \in P$ or $g^{-1} \in P$, but not both unless $g = e$. If $g, h \in P$ and a, β are the ordinals with $g \in C_a \setminus C_{a+1}$, $h \in C_\beta \setminus C_{\beta+1}$, then $gC_{a+1} \in P_a$, $hC_{\beta+1} \in P_\beta$ and, assuming e. g. $a \geq \beta$, we have $gh \in C_a \setminus C_{a+1}$ and $gh\,C_{a+1} \in P_a$, $gh \in P$. Finally, if g is as before and $x \in G$ is arbitrary, then $x^{-1}gx = g[g, x] \in gC_{a+1}$, i. e. $x^{-1}gx$ again belongs to P. Consequently, P defines a full order on G (inducing the given orders P_a on C_a / C_{a+1}).

Theorem 6 has been generalized to loops by BRUCK [1].

Corollary 7. (NEUMANN [1].) *If the lower central series of a group G terminates at e after ω steps,[5] and if the factor groups are torsion-free, then G is an O-group.*

---

[4] Such an ordinal exists, because if $g \in C_{\beta+1}$ for all $\beta < a$, then $g \in C_a$; and if $g \in C_{\beta+1}$ for all $\beta < \tau$, then $g = e$.

[5] Here ω denotes the first infinite ordinal.

Recall that the members of the lower central series of G are defined by $^0G = G$, $^{n+1}G = [G, {}^nG]$.

NEUMANN [1] proved the analogous result for the upper central series.

Theorem 8. (BIRKHOFF,[6] IWASAWA [2], NEUMANN [1].) *All free groups are O-groups.*

By a theorem of MAGNUS—WITT,[7] the lower central series of a free group satisfies the hypotheses of the preceding corollary.

BRUCK [1] notes that Theorem 8 holds for free loops. VINOGRADOV'S result mentioned at the end of Chapter III, **2** is an important generalization of Theorem 8.

Theorem 9. (IWASAWA [2], NEUMANN [1].) *Every f. o. group G is an o-epimorphic image of a f. o. free group.*

G as an abstract group may be represented as a factor group F/R of a free group F. Write $R_n = {}^nF \cap R$ where ^nF is the nth term in the lower central series of F. The groups R_n ($n = 0$, $1, \ldots$) form a descending chain with e as intersection, because $\cap R_n \subseteq \cap {}^nF = e$. Further we have

$$[F, R_n] \subseteq [F, {}^nF] \cap [F, R] \subseteq {}^{n+1}F \cap R = R_{n+1}$$

and

$$R_n/R_{n+1} = ({}^nF \cap R)/({}^{n+1}F \cap R) \cong$$
$$\cong {}^{n+1}F({}^nF \cap R)/{}_1^{n+}F \subseteq {}^nF/{}^{n+1}F.$$

Thus the factor groups R_n/R_{n+1} are torsion-free Abelian groups, on which we can define full orders P_n. Just as in the proof of Theorem 6, one can define a full order P on F on using the given full order P' of $G = F/R$ and the full orders P_n.[8] Under P, the subgroup R is convex and the natural homomorphism of F onto G is an o-epimorphism.

---

[6] BIRKHOFF has noted this result in his review in Mathematical Reviews of EVERETT—ULAM's paper [1]. This has also been noted by A. TARSKI.

[7] W. MAGNUS, *Math. Ann.*, **111** (1935), 259—280, and E. WITT, *Journ. reine u. angew. Math.*, **177** (1937), 152—160. In the case of free groups, the factor groups in the lower central series are free Abelian groups and the ωth member is e.

[8] The above method does not work if $g \notin R$, but this case is obvious.

The same inference applies to show:

Corollary 10. *Every p. o. group is an o-epimorphic image of a p. o. free group such that the kernel is f. o.*

Zaĭceva [1] and Trevisan [2] described all non-isomorphic full orders on a free Abelian group of finite rank r; there are continuously many such orders provided $r \geq 2$. Hence one gets (see Matsushita [2]): if $r \geq 2$, the free group of rank r has continuously many different full orders. Recently, Teh [1] described the full orders of torsion-free Abelian groups of finite rank.

3. The chain of convex subgroups

We wish now to investigate the system Σ of convex subgroups of a f. o. group G. This system plays a prominent part in the theory of f. o. groups. The main result is based on the following properties of Σ.

(1) Σ *is a chain containing* $\{e\}$ *and* G; *together with the subgroups* C_λ $(\lambda \in \Lambda)$ *their intersection* $\cap\, C_\lambda$ *and their union* $\cup\, C_\lambda$ *also belong to* Σ.

If C and D are convex subgroups of G and if $c \in C$, $c \notin D$ (say $c > e$), then no $d \in D$ can satisfy $e < c < d$. Consequently, $d < c$ for all $d \in D$, and so $D \subseteq C$. The other statements in (1) are obvious, and so is

(2) *If* $C \in \Sigma$ *and* $g \in G$, *then* $g^{-1} C g \in \Sigma$.

By a *jump* in Σ we mean a pair of subgroups D, C in Σ such that C contains D properly and Σ contains no subgroup between C and D; we shall use the notation $D \prec C$.[9]

(3) *If* $D \prec C$ *is a jump in* Σ, *then* D *is normal in* C *and* C/D *is isomorphic to a subgroup of the real numbers.*

Clearly, $D \prec C$ implies $a^{-1} D a \prec a^{-1} C a$ for all $a \in G$. In particular, if $a \in C$, then $a^{-1} C a = C$, and hence $a^{-1} D a = D$, i. e. D is normal in C. In view of $D \prec C$, the factor group C/D endowed with the full order induced by that of G has

---

[9] Every element $g \neq e$ of G defines a jump in Σ, for the intersection C of all convex subgroups containing g covers the union D of all other convex subgroups. Every jump arises in this way. — Σ is weakly atomic in the sense that

$$A \subset B \;(A, B \in \Sigma) \text{ implies } A \subset D \prec C \subseteq B \text{ for some } C, D \in \Sigma.$$

no non-trivial convex subgroup, thus it is o-isomorphic to a group of real numbers.

(4) *If $D \prec C$ is a jump, then* [10]

$$[N(D), N(D), C] \subseteq D.$$

Since $D \prec C$ implies $a^{-1} Da \prec a^{-1} Ca$, we obtain that if a belongs to one of $N(C)$, $N(D)$, then it belongs to the other as well, i. e. $N(C) = N(D)$. The transformation by $a \in N(D)$ leaves C and D invariant, hence it induces an automorphism $\hat{a}$ of C/D. Clearly, $\hat{a}$ preserves order, and so by (3) and Corollary 3, $\hat{a}$ commutes with all $\hat{b}$ $(b \in N(D))$. This means that $a^{-1} b^{-1} cba \equiv b^{-1} a^{-1} cab$ (mod D) or

$$[a^{-1}, b^{-1}, c] = bab^{-1} a^{-1} c^{-1} aba^{-1} b^{-1} c \in D$$

for all $a, b \in N(D)$, $c \in C$.

(5) *If $C \in \Sigma$ and $S(a)$ intersects C, then some conjugate of a lies in C.*

If the conclusion were not true, then every conjugate of a would be larger or smaller than every element of C (according as $a > e$ or $a < e$). The same would be true for the products of conjugates, contrary to hypothesis.

Now we have arrived at

Theorem 11. (RIEGER [1], PODDERYUGIN [2].) *A group G is an O-group if, and only if, it contains a system Σ of subgroups satisfying* (1)—(5).

It remains to prove the "if" part. Assume that the group G has a system Σ of subgroups such that (1)—(5) hold. Calling the jumps $D \prec C$ and $D' \prec C'$ conjugate if $C' = g^{-1} Cg$, $D' = g^{-1} Dg$ for some $g \in G$, we divide the jumps in Σ into disjoint classes of conjugacy. From each class we select a jump $D \prec C$. By (3) and (4), C/D is a torsion-free Abelian group on which the transformations by the elements of $N(D)$ commute. We may regard C/D as a module with $\Omega = N(D)$ as operator domain. By (5), C/D is Ω-torsion-free, and in view of the com-

---

[10] $N(D)$ denotes the normalizer of D in G, while $[A, B, C] = [[A, B], C]$.

mutativity of Ω and Corollary 14 of Chapter III, we can define a full Ω-order $P(C, D)$ on C/D. Let $P(g^{-1}Cg, g^{-1}Dg)$ be the induced order on $g^{-1}Cg/g^{-1}Dg$. This is compatible with the transformations by elements of $g^{-1}N(D)g = N(g^{-1}Dg)$ and does not depend on the transforming element g chosen.

Now are able to define P in G. Each $x \in G$ ($x \neq e$) defines a jump $D \prec C$ in Σ, for by (1) the intersection C of all subgroups in Σ containing x and the union D of all subgroups in Σ not containing x belong to Σ. We put $x \in P$ if, and only if, the coset xD belongs to $P(C, D)$ in C/D, and we let $e \in P$. It follows essentially in the same way as in the proof of Theorem 6 that P defines a full order on G. (The only difference is the proof of invariance which follows now from the definition of ordering for conjugate jumps.) Since the subgroups $\neq G$ in Σ are either equal to the smaller subgroup in a jump or to the intersection of such subgroups,[11] we see that the subgroups in Σ will be convex in the full order P. This completes the proof.

Conditions (1)—(5) are not yet sufficient to ensure that Σ coincides with the system of all convex subgroups in the resulting f. o. group, for the full Ω-orders $P(C, D)$ are not necessarily chosen so as to be Archimedean. (It is not known whether or not conditions (1)—(5) imply that this can be achieved.) However, if we replace (4)—(5) by a single condition (6)—of a less group-theoretic character—then it will be possible to choose the full order on G such that the given system Σ will be just the set of all convex subgroups.

For jumps $D \prec C$, let us call *principal automorphisms* the automorphisms of C/D which are transformations by elements of $N(D)$. The principal automorphisms generate in the endomorphism ring $\mathfrak{E}(C/D)$ of C/D a commutative subring $\mathfrak{R}(C/D)$ which is, by Proposition 2, isomorphic to a subring of the real number field. Since the principal automorphisms correspond to positive real numbers, if we adjoin their square roots to the field of quotients of $\mathfrak{R}(C/D)$, we get a subfield $\mathfrak{K}(C, D)$ of the real numbers.

---

[11] This is a consequence of weak atomicity; cf. footnote [9].

(6) *If $D \prec C$ is a jump in Σ, then the principal automorphisms of C/D generate a domain of integrity $\Re(C/D)$ in the endomorphism ring $\mathfrak{E}(C/D)$ of C/D, and the field $\Re(C/D)$ generated by $\Re(C/D)$ and the square roots of principal automorphisms is isomorphic to a subfield of the real numbers.*[12]

We now formulate the result:

Theorem 12. (MAL'CEV [2].) *A system Σ of subgroups of a group G is the system of all convex subgroups under some full order of G if, and only if, Σ satisfies* (1)—(3) *and* (6).

The proof of sufficiency begins with the observation that the field $\Re(C/D)$ can be identified with a subfield of the real number field, hence it is an Archimedean f. o. field. $\Re(C/D)$ will be a f. o. subring of the real numbers where the principal automorphisms are necessarily positive. We select in C/D and in the real group sets M and N, respectively, which are maximally independent with respect to $\Re(C/D)$ (or to the field of quotients of $\Re(C/D)$). Because of (3) there exists a one-to-one correspondence φ between M and a subset of N. This φ induces in the obvious manner an $\Re(C/D)$-isomorphism of C/D onto a real group K.[13] If we introduce an ordering relation into C/D by requiring that this be an *o*-isomorphism, then C/D becomes an Archimedean f. o. group such that the positive cone of C/D is invariant under multiplications by positive elements of $\Re(C/D)$. If we define on $g^{-1}Cg/g^{-1}Dg$ the induced order, the same argument as above shows that G can be f. o. so that the induced orders of the C/D are the original ones and the subgroups in Σ are convex. Moreover, no subgroup $A \notin \Sigma$ can be convex. For, by (1), A defines then a jump $D \prec C$ in Σ such that $D \subset A \subset C$. A/D is a non-trivial convex subgroup of C/D, contrary to the *o*-simplicity.

IWASAWA was the first to give a complete characterization of the system Σ of convex subgroups of a f. o. group; see his paper [2].

---

[12] MAL'CEV's original condition is somewhat weaker than (6): it requires that $\Re(C/D)$ is formally real. We have chosen the stated form in order to guarantee that Σ shall contain all convex subgroups of the arising f. o. group.

[13] By a *real group* we mean a subgroup of the additive group of the real numbers.

Corollary 13. *An O-group has a solvable normal system, thus it is an RN-group in the sense of* KUROŠ—ČERNIKOV.[14]

Corollary 14. *If the convex subgroups of a f. o. group satisfy either the minimum or the maximum condition, then every convex subgroup is normal.*

The first corollary is obvious, while the second follows from the fact that in f. o. groups the conjugates of a convex, but not normal subgroup C form an infinite chain.[15]

The order type of Σ is evidently an invariant of the f. o. group G. It is completely determined by the subset Σ_0 consisting of all "principal" convex subgroups. Here we mean by a principal convex subgroup C one generated by a single element: $C = \{a\}_\square$. A convex subgroup C is principal if, and only if, it is the greater member of a jump $D \prec C$. All other convex subgroups are unions of principal ones. If Σ is given, Σ_0 consists of the greater members of jumps. If Σ_0 is given, Σ may be obtained by adjoining to Σ_0, for every $C \in \Sigma_0$ having no immediate predecessor, the union D of all predecessors of C and then setting $D \prec C$.

4. Valuations of fully ordered Abelian groups

Now we consider Abelian groups, and we write the operation as addition.

Let A be a f. o. Abelian group and T a f. o. set with a maximal element u. A *valuation* $w(a)$ $(a \in A)$ of A is a function defined on A with values in T such that

(i) $w(a) = u$ is equivalent to $a = 0$,

(ii) $w(na) = w(a)$ for every integer $n \neq 0$,

(iii) $w(a + b) \geq \min(w(a), w(b))$.

As usual one derives from (iii), on using (ii),

(iv) if $w(a) \neq w(b)$, then $w(a + b) = \min(w(a), w(b))$.

---

[14] See e. g. KUROŠ, Теория групп (Москва, 1953), p. 367; or its English translation (New York, 1955—56), vol. 2, p. 182, where they are called SN-groups.

[15] Thus the convex subgroups are infra-invariant in the sense of M. KRASNER, *C. R. Acad. Sci. Paris*, **208** (1939), 1867—1869.

Let Π be a set indexing the set Σ_0 of principal convex subgroups of A, and inversely ordered, that is to say, if π, $\varrho \in \Pi$ and $\pi \leq \varrho$ then $C_\varrho \subseteq C_\pi$, and conversely. The *natural valuation* $v(a)$ of A is defined as

$$v(a) = \pi \quad \text{where} \quad \{a\}_\square = C_\pi.$$

Let $D_\pi \prec C_\pi$ be a jump in Σ and write $B_\pi = C_\pi/D_\pi$. The B_π are real groups. The system

$$[\Pi, B_\pi \; (\pi \in \Pi)]$$

is an invariant of A; it will be called the *skeleton* of A. If A is a subgroup of the f. o. group A' and A, A' have the same skeleton, then we say that A' is an *immediate extension* of A. If A has no proper immediate extension, it will be called *maximally valued*. More generally, if $A \subseteq A'$, there exists a natural imbedding of Π in Π'.[16] If this maps Π onto Π', we call A' an *Archimedean extension* of A, and if A has no proper Archimedean extension, it is called *Archimedean complete*. An Archimedean complete group is always maximally valued.

Since Π has maximal element μ correspoding to $0 \in \Sigma_0$, we have to agree once for all that in the skeleton always $B_\mu = 0$.

Theorem 15. (HAHN [1], RIBENBOIM [2].) *To every f. o. set Π with maximal element and to every set B_π $(\pi \in \Pi)$ of nonzero f. o. real groups there exists a f. o. Abelian group with the skeleton $[\Pi, B_\pi \; (\pi \in \Pi)]$.*

Form the lexicographic product $A = \Gamma B_\pi$ over the index set Π. If $a = \langle \ldots, b_\pi, \ldots \rangle$ with $b_\pi = 0$ for $\pi < \pi_0$ and $b_{\pi_0} \neq 0$ belongs to a convex subgroup C, then every $a' = \langle \ldots, b'_\pi, \ldots \rangle$ with $b'_\pi = 0$ for $\pi < \pi_0$ also belongs to C. Hence $\{a\}_\square = C_{\pi_0}$ consists of all vectors whose components for $\pi < \pi_0$ vanish, and the predecessor D_{π_0} of C_{π_0} consists of all vectors with 0 components for $\pi \leq \pi_0$. Consequently, C_{π_0}/D_{π_0} is o-isomorphic to B_{π_0}. The natural valuation is $v(a) = \pi_0$, and one arrives at the conclusion that the skeleton of A is the given one.

---

[16] Observe that if x, $y \in A$ generate the same convex subgroup in A, then they generate the same convex subgroup in A' too.

5. Hahn's embedding theorem

This section is devoted to the deepest result in the theory of f. o. Abelian groups. This asserts the embeddability of f. o. Abelian groups in the lexicographic product of real groups. We begin with

Lemma A. *Every f. o. Abelian group G is o-isomorphic to a subgroup of a f. o. vector space over the rational number field $\mathfrak{F}$. Among such vector spaces there exists a minimal one, V, containing G. This is unique up to o-isomorphisms over G and there exists a natural one-to-one correspondence between the sets of convex subgroups of G and of V.*

G is a torsion-free Abelian group, thus it can be embedded in a minimal divisible group V (unique up to isomorphism over G) which may be considered as a vector space over the rational numbers. Define $v \geq 0$ ($v \in V$) if, and only if, $nv \geq 0$ for some natural n with $nv \in G$. It is readily checked that this definition makes V into a f. o. vector space over $\mathfrak{F}$ and it is the only full order of V which induces that of G. Every convex subgroup C of G generates a convex subgroup C' of V such that $C' \cap G = C$. This completes the proof.

In view of this lemma, there will not be any loss of generality if *we consider only vector spaces over $\mathfrak{F}$* when examining the problem of embedding.

Let G be a f. o. vector space over $\mathfrak{F}$, $[\Pi, B_\pi (\pi \in \Pi)]$ the skeleton of G (where the real groups B_π are again f. o. vector spaces over $\mathfrak{F}$), and $v(x)$ the natural valuation of G.

The multiplication of the elements of G by arbitrary real numbers is in general not defined, but we can introduce multiplication by certain real numbers (depending on the elements) as follows. Let $x \in G$ and $v(x) = \pi$; let further $D \prec C$ be the jump in the system Σ of convex subgroups of G which corresponds to B_π. Because of the divisibility of D we may write[17] $C = B + D$ with $x \in B$ and $B \cong B_\pi$. This B can be identified

---

[17] This direct decomposition applies not only to the abstract groups, but to the order as well; for it is easy to see that this is a lexicographic sum.

with a real group, and so we can write every element of B in the form rx with a real number r. By Proposition 2, the real numbers r for which rx exists in G are uniquely determined by x and G; however, it must be emphasized that the meaning of rx may depend on the choice of B, and therefore we must think of B as fixed for a given x.

Lemma B. *If* $v(x) = v(y) = \pi$, $y \neq 0$, *there exists a unique real number* r *such that*

$$v(y - rx) > v(y).$$

Using the same notation as above, write $y = x' + z$ with $x' \in B$, $z \in D$. Then $x' = rx$ for some unique real r. Clearly, $v(z) > \pi$.

It is through the next lemma that our central objective will be attained.

Main Lemma. (HAUSNER—WENDEL [1].) *Let* $[\Pi, B_\pi \, (\pi \in \Pi)]$ *be the skeleton of the f. o. vector space*[18] G, *and for each* $\pi \in \Pi$ *let an* $e_\pi \in G$ *be selected satisfying* $v(e_\pi) = \pi$. *Let* f_π *denote the element in the lexicographic sum* $W(G)$ *of the groups* B_π *($\pi \in \Pi$) with $b_\pi \in B_\pi$ in the πth place and 0's elsewhere where* b_π *corresponds to* e_π *in* B_π. *Assume we have a proper subspace* G_0 *of* G *containing all real multiples of* e_π *that are contained in* G, *and a function* F *from* G_0 *into* $W(G)$ *such that*

(i) $F(a + b) = F(a) + F(b)$, $F(sa) = sF(a)$ *for rational numbers* s,

(ii) F *is one-to-one between* G_0 *and some subset* $F(G_0)$ *of* $W(G)$,

(iii) F *is o-preserving*,

(iv) *for every* $\pi \in \Pi$,

$$F(re_\pi) = rf_\pi$$

whenever r *is a real number such that* $re_\pi \in G_0$,

(v) *if* $f \in F(G_0)$ *and* C *is any cut in* Π, *then*

$$Cf \in F(G_0).$$

---

[18] Only vector spaces over $\mathfrak{F}$ are considered here.

[Here a *cut* C is defined by some $\pi_0 \in \Pi$ such that $Cf = g$ satisfies[19] $g(\pi) = f(\pi)$ for $\pi < \pi_0$ and $g(\pi) = 0$ for $\pi \geq \pi_0$.]

Let $x \in G$, $x \notin G_0$ and $G_1 = \{G_0, \mathfrak{F}x\}$. *Then there exists an extension of F which is a function from G_1 into $W(G)$ again satisfying* (i)—(v).

The proof consists of several steps.

(a) The set of all $v(x - y)$ with y ranging over G_0 has no maximal element. For if $v(x - y) = v(e_\pi)$, then by Lemma B there exists a real number r such that $v(x - y - re_\pi) > v(x - y)$ and here $y + re_\pi \in G_0$. We conclude that the set of values $v(x - y)$ considered has a well-ordered cofinal subset Θ indexed by ordinals a less than some limit ordinal θ. Let π_a be the elements of Θ, and let $z_a \in G_0$ such that

$$v(x - z_a) = \pi_a.$$

(b) $a < \beta$ means $v(x - z_a) < v(x - z_\beta)$ whence $v(z_a - z_\beta) = v(x - z_a) = \pi_a$. Thus $z_a - z_\beta$ and e_{π_a} are Archimedean equivalent, and by (iii) so are $F(z_a - z_\beta)$ and f_{π_a}. Therefore $F(z_a) = z'_a$ and $F(z_\beta) = z'_\beta$ $(\in W(G))$ have the same components for $\pi < \pi_a$. Define

$$x' = \langle \ldots, x'(\pi), \ldots \rangle$$

by agreeing to put $x'(\pi) = z'_a(\pi)$ if there exists a π_a greater than π, and $x'(\pi) = 0$ otherwise. Clearly, x' is well-defined and belongs to $W(G)$, for the $x'(\pi)$ vanish except on a well-ordered set. We extend F by putting $F(sx + y) = sx' + F(y)$ $(y \in G_0)$ for rational numbers s. We have to verify that the extended F fulfils (i)—(v).

(c) (i) and (iv) are obviously satisfied. To verify (v), let C be a cut at π_0 and $f \in F(G_1)$. f is of the form $f = sx' + y'$ with $y' = F(y)$, and therefore $Cf = sCx' + Cy'$. If π_0 is less than some $\pi_a \in \Theta$, then $Cx' = Cz'_a$ and so $Cf = C(sz'_a + y')$ is the cut of some element in $F(G_0)$. If π_0 exceeds all $\pi_a \in \Theta$, then $Cx' = x'$ and so $Cf = sx' + Cy' = sx' + y'_1$ for some $y_1 \in G_0$.

---

[19] $g(\pi)$ denotes the component of $g \in W(G)$ in B_π.

(d) We continue with the proof of (ii). We show that $x' = y' = F(y)$ is impossible for $y \in G_0$. If $x' = y'$ held, then $y'(\pi) = z'_a(\pi)$ for $\pi < \pi_a \in \Theta$ and therefore $v(y' - z'_a) \geq$ $\geq \pi_a = v(f_{\pi_a})$. Then $\{f_{\pi_a}\}_\square \supseteq \{y' - z'_a\}_\square$ and from (iii) we get $\{e_{\pi_a}\}_\square \supseteq \{y - z_a\}_\square$ whence $v(y - z_a) \geq v(e_{\pi_a}) = \pi_a =$ $= v(x - z_a)$. Thus $v(x - y) \geq \min(v(y - z_a), \ v(x - z_a)) =$ $= v(x - z_a)$ for all $a < \theta$. This contradicts the cofinality of Θ and the assertion in (a).

(e) Turning finally to the proof of (iii), it follows readily that it suffices to show that $x > y, x' < y'$ $(y \in G_0)$, (and similarly $x < y, \ x' > y'$), cannot hold simultaneously. Suppose, on the contrary, that they do hold, and let π_0 be the first $\pi \in \Pi$ at which $x'(\pi) \neq y'(\pi)$, i. e., $x'(\pi_0) < y'(\pi_0)$. There exists a $\pi_a \in \Theta$ such that $\pi_0 < \pi_a$, for otherwise x' would be a cut Cy' of y', and by (v), $x' = Cy' = y'_1$, contrary to (d). For such an a we have $x'(\pi_0) = z'_a(\pi_0) < y'(\pi_0)$, but $z'_a(\pi) =$ $= y'(\pi)$ for $\pi < \pi_0$. Therefore $z'_a < y'$ and $z_a < y$, since (iii) holds on G_0. We infer that $x > y > z_a, x - z_a > y - z_a >$ > 0 whence $v(y - z_a) \geq v(x - z_a) = \pi_a = v(e_{\pi_a})$. As in (d) we obtain $v(y' - z'_a) \geq \pi_a$ which means that $y'(\pi) = z'_a(\pi)$ for $\pi < \pi_a$, in particular for $\pi = \pi_0$. This contradiction completes the proof of (iii) and of the Main Lemma.

Evidently, $W(G)$ has the same skeleton as G.

Theorem 16. (HAHN's Embedding Theorem, HAHN [1].) *Every f. o. vector space G over the rational number field is o-isomorphic to a subspace of the lexicographically ordered function space*[20] $W(G)$.

If given G, take for G_0 the subspace of G spanned by the real multiples of the e_π which belong to G. It is readily verified that no sum $r_1 e_{\pi_1} + \ldots + r_k e_{\pi_k}$ with real $r_i \neq 0$ and $\pi_1 < \ldots < \pi_k$ can vanish, and therefore the function F_0 defined by (i), (iv) on G_0 satisfies all of (i)—(v). We now partially order the pairs $[G_\nu, F_\nu]$ (where G_ν is a subspace of G containing G_0 and F_ν a function from G_ν into $W(G)$ which is an extension of F_0) by putting $[G_\nu, F_\nu] \leq [G_\mu, F_\mu]$ if, and only if, $G_\nu \subseteq$

---

[20] For the definition of $W(G)$ see Main Lemma.

$\subseteq G_\mu$ and F_μ agrees with F_ν on G_ν. We can obviously apply ZORN's lemma to conclude the existence of a maximal pair $[G_\lambda, F_\lambda]$. Here G_λ coincides with G itself, for otherwise, by the Main Lemma, we could enlarge $[G_\lambda, F_\lambda]$. The proof is now complete.

We establish the following "Completeness" Theorems:

Theorem 17. *A f. o. vector space G over the rationals is maximally valued if, and only if, it is o-isomorphic to $W(G)$.*

Theorem 18. (HAHN [1].) *A f. o. Abelian group G is Archimedean complete if, and only if, in its skeleton $[\Pi, B_\pi (\pi \in \Pi)]$ all the groups B_π are o-isomorphic to the whole real group and G is o-isomorphic to $W(G)$.*

The two proofs are similar, we confine ourselves to giving the first one in detail. If G is not o-isomorphic to $W(G)$, then it is o-isomorphic to a proper subspace of $W(G)$, and so it is not maximally valued. Conversely, if $G \cong_o W(G)$ and if there were an immediate extension H of G, then (supposing H to be again a vector space) we can apply the Main Lemma with F the given o-isomorphism of G onto $W(G)$ and with some $x \in H$ not in G. If F_1 is the extension of F to $G_1 = \{G, \mathfrak{F}x\}$, then $F_1(x) = x'$ belongs to $W(H) = W(G)$. Hence there exists a $y \in G$ such that $F_1(y) = F_1(x)$ which is impossible in G_1.

The original proof of HAHN was extremely long and complicated. Recently, several authors have obtained simpler proofs and generalizations. The proof above is based on an idea of HAUSNER—WENDEL [1]: they proved HAHN's theorem for vector spaces over the real field and CLIFFORD [4] observed that their method works in the general case as well. For other proofs see BANASCHEWSKI [1], GRAVETT [2], RIBENBOIM [2], CONRAD [1], [7]. The last author has extended the theorem to certain p. o. Abelian groups and to even more general systems; he uses decompositions of the given group.

Recently, P. CONRAD, J. HARVEY and CH. HOLLAND proved HAHN's embedding theorem for commutative l. o. groups.

The results of this section will continue to hold if we replace the underlying field $\mathfrak{F}$ by the real number field. Moreover, if the dimension is at most countable, we can then prove somewhat more:

Theorem 19. (ERDŐS [1].) *Let V be a f. o. vector space over the real numbers of countable dimension. Then V has a f. o. basis*

$b_1, \ldots, b_n, \ldots$ (with $n \leq \dim V$) such that

$$r_1 b_1 + \ldots + r_n b_n > 0 \qquad (r_i \text{ real numbers}, r_i \neq 0)$$

if, and only if, the coefficient of the greatest b_i in the ordering is positive.

The vector space V has a basis $a_1, \ldots, a_n, \ldots$ over the real numbers. Put $b_1 = |a_1|$, and assume that $b_1, \ldots, b_n \in V$ have been selected such that (1) $b_i > 0$, (2) the values $v(b_i)$ for the natural valuation v are different, and (3) the vectors $b_1, \ldots, b_n$ span the same subspace V_n as $a_1, \ldots, a_n$. Consider a_{n+1}; this is independent of V_n. If $v(a_{n+1}) \neq v(b_i)$ for $i = 1, \ldots, n$, we set $b_{n+1} = |a_{n+1}|$, while if $v(a_{n+1})$ equals some $v(b_i)$, then let $v(b_j)$ be the greatest $v(b_i)$ for which there exists a linear combination $x = r_1 b_1 + \ldots + r_n b_n + r a_{n+1}$ with $r \neq 0$ and $v(x) = v(b_j)$. By Lemma B, there exists a real number r' such that $v(x - r' b_j) > v(b_j)$, and we define $b_{n+1} = |x - r' b_j|$. Then we have $n + 1$ vectors $b_1, \ldots, b_n, b_{n+1}$ with the same properties (1)–(3). So we can construct a new basis $b_1, \ldots, b_n, \ldots$ with positive b_n such that the values $v(b_n)$ are all different. These b_n fulfil the requirements, for by property (iv) of valuations (see **4**) we have

$$v(r_1 b_{i_1} + \ldots + r_k b_{i_k}) = \min_j v(r_j b_{i_j}) = \min_j v(b_{i_j}),$$

and for $v(x) > v(y)$, $x + y > 0$ holds if, and only if, $y > 0$. Q. E. D.

Let us note that the f. o. vector spaces over the real numbers of dimension $\leq \aleph_0$ are uniquely determined by their skeletons $[\Pi, B_\pi (\pi \in \Pi)]$, moreover, by Π alone, because all the B_π are o-isomorphic to the whole real group. It is readily seen that the order type of Π may be any countable order type.

Methods of ordering vector spaces over fields have been discussed by Conrad [4], [5].

6. Cyclically ordered groups

Now we turn to the rather special problem of cyclical ordering. This differs essentially from the order relation studied so far in that it is a ternary relation. We shall show that every cyclically ordered group may be obtained from some f. o. group by a special construction.

A group K is said to be *cyclically ordered*[21] if for some triplets a, b, c of different elements of K a relation $(a, b. c)$ is defined with the following properties:

C1. exactly one of (a, b, c) and (a, c, b) holds;

C2. (a, b, c) implies (b, c, a);

C3. (a, b, c) and (a, c, d) imply (a, b, d);

C4. (a, b, c) implies (xa, xb, xc) and (ay, by, cy) for all $x, y \in K$.

If in C3 we change the roles of a and c, b and d, we obtain that (a, b, c) and (a, c, d) imply (b, c, d). In this case each of the four cyclic order relations holds which is got from (a, b, c, d) by omitting an element. In general, if $(a_1, a_2, \ldots, a_n)$ $(n \geq 3)$ means that (a_i, a_j, a_k) holds for all $i < j < k$, then by making use of C2 and C3 it is easy to show by induction:

Lemma. *If*

$$(a, b_i, b_{i+1}) \quad (i = 1, 2, \ldots, n-1),$$

then

$$(a, b_1, b_2, \ldots, b_n).$$

In particular, $a, b_1, b_2, \ldots, b_n$ are different.

The following examples are of importance.

1. A cyclic group $K = \{a\}$ of order n may be cyclically ordered by setting (a^k, a^l, a^m) $(0 \leq k, l, m \leq n-1)$ exactly if $k < l < m$ or $l < m < k$ or else $m < k < l$. A different generator a' of K gives rise to a different cyclical order.

2. The complex numbers on the unit circle (or an arbitrary subgroup) form a cyclically ordered group if (a, b, c) means that the numbers a, b, c follow counter-clockwise.

3. If G is any f. o. group, then a cyclical order may be defined on G by the rule:

$$(a, b, c) \text{ if either } a < b < c \text{ or } b < c < a \text{ or } c < a < b.$$

This may be called the induced cyclical order.

[21] See RIEGER [1].

4. Let G be a f. o. group containing in the centre an element $z > e$ such that $\{z\}_\square = G$. Then the factor group $G/\{z\} = K$ can be cyclically ordered by putting (a, b, c) for cosets a, b, c mod $\{z\}$ if (r_a, r_b, r_c) holds in the sense of the foregoing example in G for the unique representatives r_a, r_b, r_c of the cosets a, b, c with $e \leq r_a, r_b, r_c < z$.

The cyclically ordered torsion groups can be described explicitly. The same holds for the sets of periodic elements of cyclically ordered groups.

Theorem 20. *In a cyclically ordered group K the elements of finite order lie in the centre of K. They form a subgroup T of K o-isomorphic to a subgroup of the cyclically ordered group C of complex roots of unity.*

Assume that (e, b, a) for some $a \in K$ of finite order n. If $(e, a^{-1}ba, b)$, then by C4 also $(e, a^{-k-1}ba^{k+1}, a^{-k}ba^k)$ for $k = 0, 1, \ldots, n-1$ which is impossible because of the lemma. If $(e, a^{-1}ba, b)$ is not true and $a^{-1}ba \neq b$, then (e, aba^{-1}, b) holds and we get a similar contradiction. Thus $a^{-1}ba = b$ for all $b \in K$ with (e, b, a). The only remaining case of interest is when (e, a, b). Then, by C4, $(e, a^{-1}b, a^{-1})$ holds and what has been proved implies that $a^{-1}b$ and a^{-1} commute, hence so do a and b. Thus a belongs to the centre, and the elements of finite order form a subgroup T of K.

Next we verify that every finite ($=$ finitely generated) subgroup $A \neq e$ of T is cyclic. A surely contains an element $a \neq e$ such that no $x \in A$ satisfies (e, x, a), for otherwise we could construct a sequence $a, a_2, a_3, \ldots$ with (e, a_{i+1}, a_i) which contradicts the lemma, in view of the finiteness of A. If this a is of order 2, then (e, a, y) for some $y \in A$ implies (a, e, ay) which is absurd; thus $A = \{a\}$. If a is of order $n \geq 3$, then (e, a, a^2), and so we get (a^k, a^{k+1}, a^{k+2}) for every k. It is impossible to have (a^k, x, a^{k+1}) for any $x \in A$, since then (e, xa^{-k}, a). Also (a^k, a^{k+1}, x) for some $x \in A$ and for every k contradicts the lemma. Hence $A = \{a\}$ and $(e, a, a^2, \ldots, a^{n-1})$. Consequently A is a locally cyclic periodic group and therefore it is isomorphic to a subgroup of C.

T can be obtained as the union of an ascending chain of

finite groups A_n; $T = \bigcup A_n$. Every A_n is cyclic (of order m_n) and the argument of the preceding paragraph shows that it has a uniquely determined generator a_n such that $(e, a_n, a_n^2, \ldots, a_n^{m_n-1})$. In this case the mapping $a_n^k \to \exp 2k\pi i/m_n$ is an o-isomorphism φ_n of A_n into C. Since this is clearly the only o-isomorphism of A_n into C, we infer that φ_{n_1} and φ_{n_2} $(n_1 < n_2)$ agree on A_{n_1}. Therefore the φ_n define an o-isomorphism φ of T into C. Q. E. D.

The next result shows that the construction of Example 4 is the most general one.

Theorem 21. (RIEGER [1].) *To every cyclically ordered group K there exists a fully ordered group G and an element z in the centre of G such that K may be obtained from G as described in Example* 4.

We define the group G as a central Schreier extension of an infinite cyclic group $\{z\}$ by the given group K. Let G consist of all pairs $\langle z^k, a \rangle$ with integers k and $a \in K$ subject to the rules:

$$\langle z^k, a \rangle = \langle z^l, b \rangle \text{ if, and only if, } k = l \text{ and } a = b,$$

and

$$\langle z^k, a \rangle \cdot \langle z^l, b \rangle = \langle z^{k+l} f_{a,b}, ab \rangle$$

where the factors $f_{a,b}$ are defined by

$$f_{a,b} = \begin{cases} e \text{ if } a = e \text{ or } b = e \text{ or } (e, a, ab), \\ z \text{ if } ab = e \text{ (with } a \neq e) \text{ or } (e, ab, a). \end{cases}$$

A straightforward calculation shows that the associativity conditions $f_{a,b}\, f_{ab,c} = f_{a,bc}\, f_{b,c}$ (for all $a, b, c \in K$) are satisfied. Thus G exists and we let the positive cone P of G consist of all $\langle z^k, a \rangle$ with $k \geq 0$. Then P satisfies the conditions of Theorem 2 in Ch. II, hence it defines a partial, moreover a full order on G. Now $\langle z, e \rangle$ lies in the centre of G and obviously $\{\langle z, e \rangle\}_\square = G$, $G/\{\langle z, e \rangle\} \simeq K$. Every coset $\bar{a}$ mod $\{\langle z, e \rangle\}$ may be represented by a unique element of the form $\langle e, a \rangle$ and—according to the prescription of Example 4—we have to put $(\bar{e}, \bar{a}, \bar{b})$ $(a \neq e \neq b \neq a)$ if, and only if, $\langle e, a \rangle < \langle e, b \rangle$

in G, or what amounts to the same, $\langle z^{-1}, a^{-1} \rangle \langle e, b \rangle =$
$= \langle z^{-1} f_{a^{-1},b}, \ a^{-1} b \rangle \in P$. This happens exactly in case
$(e, a^{-1}b, a^{-1})$ holds, that is to say, (e, a, b). This completes
the proof.

Note that G contains a maximal convex subgroup H not containing
$\langle z, e \rangle$ and G/H must be o-isomorphic to a real group. Since
$H \cap \{\langle z, e \rangle\} = e$, H can be looked upon as a subgroup of K. It can
be characterized in K as the set of all $h \in K$ for which either $(e, h,$
$h^2, \ldots, h^n, \ldots)$ or $(e, h^{-1}, h^{-2}, \ldots, h^{-n}, \ldots)$ $(n = 1, 2, \ldots)$. (Its
cyclical order is induced by a full order.) Thus a cyclically ordered group
is a Schreier extension of a f. o. group by a subgroup of the real numbers
mod 1. (Cf. Świerczkowski [1].)

Shepperd [1], [2], [3] studied groups where a ternary relation
"betweenness" or a quaternary relation "separation" is defined having similar properties as the corresponding concepts known for the
points of the real line, together with the monotony law.

CHAPTER V

LATTICE-ORDERED GROUPS

1. Algebraic rules

This section is devoted to establishing some basic properties of l. o. groups which we shall use a number of times.[1]

Recall the definition: a l. o. group is a set G satisfying

L1. G is a group under multiplication,

L2. G is a lattice under a relation $\leq$,

L3. $a \leq b$ implies $ca \leq cb$ and $ac \leq bc$ for all $c \in G$.

In Chapter II, 1 it has been shown that L3 has several equivalent formulations. We can now add to them the following two:

A) L3 is equivalent to either one of the identities (STONE [1]):

$$c(a \vee b)d = cad \vee cbd \quad \text{for all } a, b, c, d \in G,$$

$$c(a \wedge b)d = cad \wedge cbd \quad \text{for all } a, b, c, d \in G.$$

Either of these trivially implies L3, while from L3 it follows easily that $x \geq cad \vee cbd$ is equivalent to $x \geq c(a \vee b)d$, and dually.

Thus *the class of l. o. groups is equationally definable.*[2]

B) For all $a, b, c, d \in G$ we have

$$c(a \wedge b)^{-1} d = ca^{-1} d \vee cb^{-1} d,$$

$$c(a \vee b)^{-1} d = ca^{-1} d \wedge cb^{-1} d.$$

In Chapter II we have proved these identities for $c = d = e$; the general case is an immediate consequence of A).

---

[1] The general theory of l. o. groups is due to BIRKHOFF [1].

[2] A class of algebraic systems which can be characterized by identities containing finitary operations defined for all (finite) sets of elements is called *equationally definable*. Such a class is closed under forming subsystems, homomorphic images and complete direct products. An important result of BIRKHOFF's states that an equationally definable algebraic system may be represented as a subdirect union of subdirectly irreducible systems of the same kind.

C) In the special case $c = a$, $d = b$. B) gives

$$a(a \wedge b)^{-1} b = a \vee b \quad \text{for all } a, b \in G.$$

If G is commutative, we obtain

$$ab = (a \wedge b)(a \vee b) \quad \text{for all } a, b \in G.$$

D) *A l. o. group is a distributive lattice.* By a known criterion of distributivity, it suffices to verify that $a \vee x = a \vee y$ and $a \wedge x = a \wedge y$ imply $x = y$. By making use of C), we have

$$x = (a \wedge x)a^{-1} a(a \wedge x)^{-1} x = (a \wedge x)a^{-1}(a \vee x) =$$
$$= (a \wedge y)a^{-1}(a \vee y) = y.$$

E) *Lattice-order is isolated, hence a l. o. group G is torsion-free.* Assume that $a \in G$ satisfies $a^n \geq e$ for a natural integer n. Then, by A),

$$(a \wedge e)^n = a^n \wedge \ldots \wedge a \wedge e = a^{n-1} \wedge \ldots \wedge a \wedge e = (a \wedge e)^{n-1}.$$

Hence $a \wedge e = e$ and $a \geq e$.

EVERETT and ULAM [1] have shown that in l. o. groups $a^n > b^n$ ($> e$) may hold for some natural integer n without the validity of $a > b$. E) implies that this cannot happen in the commutative case.

Strong isolatedness does not hold in general—as may be shown by **Example 10 in Chapter II, 3**.

F) *A p. o. group G is a l. o. group if (and only if), for every $a \in G$, the l. u. b. $a \vee e$ exists in G.*[3] If $a \vee e$ exists for every a, then define $a \vee b = (ab^{-1} \vee e)b$ and $a \wedge b = (a^{-1} \vee b^{-1})^{-1}$ which are easily proved to be the l. u. b. and the g. l. b. for a and b.

G) *In l. o. groups, complete integral closure and the Archimedean property are equivalent.* (BIRKHOFF [1].) Since the first property implies the second one in any p. o. group, we establish only the inverse implication. Assume that $a^n \leq b$ $(n = 1, 2, \ldots)$ in a l. o. group G. Then

$$(a \vee e)^n = a^n \vee a^{n-1} \vee \ldots \vee a \vee e \leq b \vee e \quad \text{for } n = 1, 2, \ldots$$

---

[3] For a definition of l. o. groups in terms of $a \vee e$ see MICHIURA [1] and LINÉS ESCARDÓ—MALLOL BALMAÑA [1].

and evidently

$$(a \vee e)^{-n} \leq e \leq b \vee e \quad \text{for} \quad n = 1, 2, \ldots.$$

Thus $(a \vee e)^n \leq b \vee e$ for all integers n. Therefore, by the Archimedean property, $a \vee e \leq e$, that is, $a \leq e$. Q. E. D.

Observe that an Archimedean l. o. group may have a non-Archimedean homomorphic image. This is shown by the l. o. group of all real functions on the positive real numbers whose factor group with respect to the bounded functions is a non-Archimedean l. o. group. The function x^2 is positive and infinitely small relative to x^4.

H) For l. o. groups the following refinement theorem holds.

Theorem 1. (RIESZ [1], BIRKHOFF [1], LORENZEN [3].) *Let G be a l. o. group and a_i, b_j elements of G such that*

$$a_1 a_2 \ldots a_m = b_1 b_2 \ldots b_n \quad (a_i \geq e, \ b_j \geq e).$$

There exist elements c_i in G $(i = 1, \ldots, m; \ j = 1, \ldots, n)$ satisfying

a) $c_{ij} \geq e$,

β) $a_i = c_{i1} c_{i2} \ldots c_{in} \quad (i = 1, \ldots, m)$,

γ) $b_j = c_{1j} c_{2j} \ldots c_{mj} \quad (j = 1, \ldots, n)$.

Moreover, we may assume that

δ) $c_{i+1,j} \ldots c_{mj} \wedge c_{i,j+1} \ldots c_{in} = e$ *for all* $i < m, \ j < n$,

and under this assumption the c_{ij} are uniquely determined.

We begin with the case $m = n = 2$, $a_1 a_2 = b_1 b_2$. Define

$$c_{11} = a_1 \wedge b_1, \ c_{12} = c_{11}^{-1} a_1, \ c_{21} = c_{11}^{-1} b_1, \ c_{22} = a_2 \wedge b_2.$$

Then we have $c_{21} = (a_1^{-1} \vee b_1^{-1}) b_1 = a_1^{-1} b_1 \vee e = a_2 b_2^{-1} \vee e = a_2 c_{22}^{-1}$, and analogously $c_{12} = b_2 c_{22}^{-1}$ whence β) and γ) are satisfied. δ) also holds, for $c_{12} \wedge c_{21} = c_{11}^{-1} (a_1 \wedge b_1) = e$.

Suppose the theorem true for integers m', n' with $m' \leq m$, $n' < n$ $(n \geq 3)$, and write the given equality in the form

$$a_1 a_2 \ldots a_m = b_1 b_2 \ldots (b_{n-1} b_n).$$

By induction hypothesis, there exist positive elements c_{ij} $(i = 1, \ldots, m; \ j = 1, \ldots, n-2)$ and d_i $(i = 1, \ldots, m)$

such that

$$a_i = c_{i1} \ldots c_{i,n-2}\, d_i$$

and

$$b_j = c_{1j} \ldots c_{mj} \quad (j \leq n - 2), \quad b_{n-1}\, b_n = d_1 \ldots d_m.$$

We again apply the induction hypothesis, now to the last equation, to conclude the existence of elements $c_{ij} \geq e$ ($i = 1, \ldots, m;\ j = n - 1, n$) with the property

$$d_i = c_{i,n-1}\, c_{in}, \quad b_j = c_{1j} \ldots c_{mj}.$$

We may in addition assume δ); then the induction hypothesis yields

$$c_{i+1,j} \ldots c_{mj} \wedge c_{i,j+1} \ldots c_{i,n-2}\, d_i = e \text{ for all } i < n,\ j \leq n - 2,$$

and

$$c_{i+1,n-1} \ldots c_{m,n-1} \wedge c_{in} = e,$$

respectively, thus establishing δ).

It remains to verify the unicity assertion. By multiplication by c_{ij} we obtain from δ) the equality

$$c_{ij}\, c_{i+1,j} \ldots c_{mj} \wedge c_{ij}\, c_{i,j+1} \ldots c_{in} = c_{ij}$$

where

$$c_{ij}\, c_{i+1,j} \ldots c_{mj} = (c_{1j} \ldots c_{i-1,j})^{-1}\, b_j$$

and

$$c_{ij}\, c_{i,j+1} \ldots c_{in} = (c_{i1} \ldots c_{i,j-1})^{-1}\, a_i.$$

This shows that the c_{ij} may be computed recursively, completing the proof.

Corollary 2. *If* $a, b_1, \ldots, b_n$ *are positive elements of a l. o. group* G *such that*

$$a \leq b_1 \ldots b_n,$$

then there exist positive elements $a_1, \ldots, a_n$ *in* G *satisfying*

$$a = a_1 \ldots a_n \text{ with } a_j \leq b_j \ (j = 1, \ldots, n).$$

There exists an $a' \geq e$ with $aa' = b_1 \ldots b_n$ and we apply the preceding theorem to this decomposition.

2. Orthogonality

The elements a, b of a l. o. group G are called *orthogonal*,[4] in sign:

$$a \perp b,$$

if $a \wedge b = e$. Clearly, $x \wedge y = z$ is equivalent to the orthogonality of xz^{-1} and yz^{-1}.

Let us consider some simple properties of orthogonal elements.

A) *If $a \perp b$ and $c \geq e$, then $a \wedge bc = a \wedge c$ (and $a \wedge cb = a \wedge c$).* In fact, $a \wedge c = a \wedge (a \wedge b)c = a \wedge ac \wedge bc = a \wedge bc$, since $ac \geq a$. (Thus $a \leq bc$, $c \geq e$ and $a \perp b$ imply $a \leq c$.)

B) *$a \perp b$ and $a \perp c$ imply $a \perp bc$.* This is an immediate consequence of A).

C) *$a \perp b$ and $a \perp c$ imply $a \perp b \wedge c$ and $a \perp b \vee c$.* The first orthogonality relation is trivial, while the second is a consequence of distributivity: $a \wedge (b \vee c) = (a \wedge b) \vee (a \wedge c) = e$.

It follows that the set D_a consisting of all elements of G orthogonal to a is a subsemigroup and a convex sublattice of the positive cone of G.

D) *If $a_1, \ldots, a_n$ are pair-wise orthogonal elements, then*

$$a_1 \vee \ldots \vee a_n = a_1 \ldots a_n.$$

By B), $a_1 \ldots a_{n-1}$ is orthogonal to a_n, thus C) in 1 implies $(a_1 \ldots a_{n-1}) \vee a_n = a_1 \ldots a_{n-1} a_n$. If we assume—as a basis of induction—that $a_1 \vee \ldots \vee a_{n-1} = a_1 \ldots a_{n-1}$, then we arrive at once at the desired equality.

E) In the special case $n = 2$ we get from D) that $a_1 a_2 = a_1 \vee a_2 = a_2 a_1$ whenever $a_1 \perp a_2$, i. e. *orthogonal elements commute.*

KUTYEV [1] has shown that orthogonality can be defined more generally, in arbitrary p. o. groups G. If P is the positive cone of G, then call $a, b \in G$ orthogonal if $Pa^{-1} \cap Pb^{-1} = P$. It follows at once that in l. o. groups this definition yields the same concept as that above.

---

[4] Another terminology is *disjoint*.

Call an element $p \in G$ $(p > e)$ an *atom* if

$$e \leq x < p \quad \text{implies} \quad x = e.$$

Obviously, if p is an atom and a is an arbitrary element $(> e)$ of G, then either $p \perp a$ or $p \leq a$.

F) *A positive element* p $(\neq e)$ *is an atom if, and only if, it has the property that*

$$p \leq ab \, (a, b \geq e) \quad \text{implies} \quad p \leq a \text{ or } p \leq b.$$

If an atom p satisfies $p \leq ab$, but neither $p \leq a$ nor $p \leq b$, then $p \perp a$ and $p \perp b$ whence $p \perp ab$, a contradiction. Conversely, if p $(> e)$ has the stated property and $e \leq a < p$, then for $b = a^{-1}p$ we have $p \leq ab$ whence $p \leq b$, and so $a = e$.

G) *If* $[p_a]_{a \in A}$ *is a set of (distinct) atoms of G, then the subgroup G' they generate is a free Abelian group with the p_a as free generators.* Different atoms are orthogonal to each other, therefore, by E), the p_a commute. Assume that $p_1^{n_1} \ldots p_t^{n_t} \geq e$ for a finite number of atoms $p_1, \ldots, p_t$. We can write this inequality, without loss of generality, in the form $p_1^{n_1} \ldots p_r^{n_r} \geq p_{r+1}^{-n_{r+1}} \ldots p_t^{-n_t}$ where only non-negative exponents occur. From B) we conclude that the two sides are orthogonal, thus the right member is e, and every $n_i \geq 0$. Hence $p_1^{n_1} \ldots p_t^{n_t} = e$ implies $n_i = 0$, as we wished to prove.

We have also shown that G' is the direct product of the f. o. groups $\{p_a\}$.

H) We shall say that the l. o. group G satisfies the minimum condition if every non-empty set of positive elements of G contains a minimal member.

Theorem 3. (WARD [1], BIRKHOFF [1].) *A l. o. group G is the direct product of f. o. cyclic groups if, and only if, it satisfies the minimum condition.*

The "only if" part being obvious, let us suppose that G is a l. o. group with minimum condition. For any $a \in G$, $a > e$, the set $[x \in G \mid e < x \leq a]$ contains a minimal member p_1 which is clearly an atom. If $a_1 = p_1^{-1} a > e$, we obtain again an atom p_2 satisfying $e < p_2 \leq a_1$. If $a_2 = p_2^{-1} a_1 > e$, we continue this process until we arrive at some $a_k = p_k^{-1} a_{k-1} = e$. Since $a > a_1 > a_2 > \ldots$, the minimum condition ensures the existence of such a k. Then $a = p_1 p_2 \ldots p_k$ is a product of atoms, so that the subgroup G' mentioned in G) coincides with G. An application of G) concludes the proof.

An analogue of Theorem 3 for l. o. semigroups has been proved by DUBREIL-JACOTIN [2]; cf. also DUBREIL [2].

3. Carriers

By using orthogonality, we can divide the positive elements of a l. o. group G into disjoint classes. We put two positive elements in the same class if, and only if, the same elements of $P = P(G)$ are orthogonal to them. The classes are called *carriers* (or *filets*).[5] Thus if a subset C of P is a carrier, then $a, b \in C$ and $a \perp x$ imply $b \perp x$. The carrier containing a will be denoted by $a^\frown$. The carrier $e^\frown$ consists of e alone and in a l. o. group with minimum condition $a^\frown$ consists of those elements $b \in P$ which contain the same atoms as a (but no others).

The following fact is readily deduced from B) and C) of **2**:

Proposition 4. (JAFFARD [1].) *A carrier is a subsemigroup and a convex sublattice of P.*

The set $\mathfrak{C}$ of all carriers of G can be p. o. by putting

$$a^\frown \geq b^\frown \text{ if, and only if, } a \perp x \text{ implies } b \perp x.$$

It is evident that this definition is independent of the representatives a, b of $a^\frown$, $b^\frown$. It is also clear that the mapping $a \to a^\frown$ is isotone.[6]

We intend to show that $(a \wedge b)^\frown$ and $(a \vee b)^\frown$ are the g. l. b. and the l. u. b., respectively, of $a^\frown$ and $b^\frown$. Obviously, $(a \wedge b)^\frown \leq$ $\leq a^\frown$ and $b^\frown$. Assume that $c^\frown \leq a^\frown$ and $b^\frown$. If $a \wedge b \perp x$, then $a \perp b \wedge x$; hence $c \perp b \wedge x$, $b \perp c \wedge x$, and so $c \perp c \wedge x$, $c \perp x$. Therefore, $c^\frown \leq (a \wedge b)^\frown$, establishing the first assertion.

---

[5] This notion has been introduced by JAFFARD [1]. In a subsequent paper [6] he considers carriers in certain p. o. Abelian groups which are more general than l. o. groups.

[6] JAFFARD [1] has shown that a minimal carrier $\neq e^\frown$ generates a f. o. subgroup of G. Cf. Lemma B in **6**.

In the second case $a^\wedge, b^\wedge \leq (a \vee b)^\wedge$ is trivial, while if $c^\wedge \geq$ $\geq a^\wedge, b^\wedge$, then $c \perp x$ implies $a \perp x$ and $b \perp x$ whence, by C), $a \vee b \perp x$, consequently, $c^\wedge \geq (a \vee b)^\wedge$. Thus $\mathfrak{C}$ is a lattice in which

$$a^\wedge \wedge b^\wedge = (a \wedge b)^\wedge \quad \text{and} \quad a^\wedge \vee b^\wedge = (a \vee b)^\wedge.$$

By 1 D), $\mathfrak{C}$ is a distributive lattice. The equality $a^\wedge \vee b^\wedge =$ $= (ab)^\wedge$ may be proved similarly. This proves the first part of

Theorem 5.[7] (JAFFARD [6], PIERCE [1].) *The mapping φ: $a \to a^\wedge$ from the positive cone P of a l. o. group G onto the distributive lattice $\mathfrak{C}$ of all carriers of G is a lattice homomorphism with kernel e, satisfying*

$$(1) \qquad\qquad (ab)^\wedge = a^\wedge \vee b^\wedge.$$

φ may be characterized as the maximal[8] lattice homomorphism of P whose kernel is e.

Let ψ be any lattice homomorphism of P with kernel e. Define an equivalence relation ϱ by taking $a, b \in P$ equivalent under ϱ if, and only if, $\psi(a) \wedge \psi(x) = \psi(e)$ is equivalent to $\psi(b) \wedge \psi(x) = \psi(e)$. It follows easily that ϱ is a congruence relation of P. Denote by η the homomorphism of P onto the equivalence classes of ϱ. Since the kernels are equal to e, $a \wedge x = e$ is equivalent to $\psi(a) \wedge \psi(x) = \psi(e)$ which is equivalent to $\eta(a) \wedge \eta(x) = \eta(e)$. The same holds for b, therefore $a^\wedge = b^\wedge$ if, and only if, $\eta(a) = \eta(b)$. Thus η is essentially the same as φ. Since η is greater than or equal to ψ, the statement follows.

PIERCE [1] also proved that φ is the only lattice homomorphism of P whose kernel is e and whose image is disjunctive in the sense that to any two different elements there exists a third orthogonal to just one of the given two. GOFFMAN [1] has shown that φ preserves suprema (i. e., l. u. b. of any subsets, if they exist) and in case G is an Archimedean l. o. group, then φ is the unique supremum preserving lattice homomorphism of P whose kernel is e and which satisfies (1). A counterexample shows that this fails to hold in the non-Archimedean case.

For the lattice $\mathfrak{C}$ of carriers in commutative l. o. groups see RIBENBOIM [3].

---

[7] The first assertion is due to JAFFARD, the second to PIERCE.

[8] A homomorphism ψ is called greater than another one χ if every class under χ is contained in some class under ψ.

We shall say that the carriers of G are *invariant* if

$$a^\wedge = (x^{-1} ax)^\wedge \quad \text{for all } a, x \in G,$$

that is, if $a \perp b$ implies $a \perp x^{-1} bx$.

Proposition 6. (KONTOROVIČ—KUTYEV [1].) *The carriers of a l. o. group G are invariant if, and only if,*

$$a \perp x^{-1} ax \quad \text{implies} \quad a = e.$$

Assume the carriers of G are invariant and $a \perp x^{-1} ax$. Then $a \perp a$ and $a = e$. Conversely, if $a \perp x^{-1} ax$ implies $a = e$, and if $a \perp b$, then also $a \wedge x^{-1} bx \perp xax^{-1} \wedge b$ whence $a \wedge x^{-1} bx = e$, $a \perp x^{-1} bx$, that is, the carriers are invariant.

Proposition 7. *The class of l. o. groups with invariant carriers is equationally definable.*

The orthogonality of a and b is equivalent to the existence of an x satisfying $a = x \vee e$ and $b = x^{-1} \vee e$. In fact, if $a \perp b$, then $x = ab^{-1}$ has the required property, while if a, b have the indicated form, then $a \perp b$ (see the next section F)). Consequently, the invariance of the carriers may be expressed as $x \vee e \perp y^{-1}(x^{-1} \vee e)y$, that is,

$$(x \wedge y^{-1} x^{-1} y) \vee e = e \quad \text{for all } x, y \in G.$$

As a corollary we obtain that every l. o. subgroup and factor group of a l. o. group with invariant carriers has the same property.

Example 10 in Chapter II, **3** exhibits a l. o. group in which the carriers are not invariant (with $x = a^{-3} b^5$ and $y = c$ we have $(x \wedge y^{-1} x^{-1} y) \vee e = a^{-5} b^3 \vee e = b^3 > e$).

The concept of carrier has been generalized by KONTOROVIČ and KUTYEV [1] to arbitrary p. o. groups. It is based on the generalized notion of orthogonality mentioned in **2**. Moreover, carriers relative to a subsemigroup of the positive cone are also studied.

4. Positive and negative parts; absolutes

The following concepts are of great importance in the theory of l. o. groups.

The *positive part*[9] a^+, the *negative part* a^- and the *absolute*

---

[9] This concept has been used first by RIESZ [1] in a special case; the general concept is due to BIRKHOFF [1]. For the absolute see KANTOROVITCH [1].

$|a|$ of an element a are defined as the elements

$$a^+ = a \vee e, \qquad a^- = a \wedge e, \qquad |a| = a \vee a^{-1}.$$

They have the following elementary properties:

A) $a^+ \geq e$ and $a^- \leq e$. *Equality holds if, and only if, $a \leq e$ and $a \geq e$, respectively.*

B) $(a^{-1})^+ = (a^-)^{-1}$, for $a^{-1} \vee e = (a \wedge e)^{-1}$. Similarly, $(a^{-1})^- = (a^+)^{-1}$.

C) $(ab)^+ \leq a^+ b^+$ and $(ab)^- \geq a^- b^-$, because

$$(ab)^+ = ab \vee e \leq ab \vee a \vee b \vee e = (a \vee e)(b \vee e) = a^+ b^+,$$

and dually.

D) $(a^n)^+ = (a^+)^n$ and $(a^n)^- = (a^-)^n$ for $n > 0$. In fact, if $0 \leq k \leq n$, then

$$(a^{n-k} \vee a^{-k})^n = a^{(n-k)n} \vee \ldots \vee a^{(n-k)k} a^{-k(n-k)} \vee \ldots \geq e$$

whence, by isolatedness, $a^{n-k} \vee a^{-k} \geq e$ and $a^n \vee e \geq a^k$. Therefore

$$(a^+)^n = (a \vee e)^n = a^n \vee \ldots \vee a^k \vee \ldots \vee e = a^n \vee e = (a^n)^+.$$

The dual argument or B) proves the second identity.—Hence it is easy to conclude that in a commutative l. o. group we have $(a \vee b)^n = a^n \vee b^n$ and $(a \wedge b)^n = a^n \wedge b^n$ for every positive n; in fact, $(a \vee b)^n = ((ab^{-1})^+)^n b^n = ((ab^{-1})^n)^+ b^n = a^n \vee b^n$, and dually.

E) $a = a^+ a^- = a^- a^+$, for $a(a^-)^{-1} = a(a^{-1} \vee e) = e \vee a = a^+$, and similarly for the other side. In particular, it follows that a^+ and a^- commute.

F) $a^+ \perp (a^-)^{-1}$, because $a^+ \wedge (a^-)^{-1} = (a \wedge e)(a^-)^{-1} = e$.

From E) and F) we see that a^+ and $(a^-)^{-1}$ may be looked upon as the numerator and denominator, respectively, of a in its "lowest terms", in the following sense: if $a = xy^{-1}$ with $x \geq e$, $y \geq e$, then there exists an element $z \geq e$ such that $x = a^+ z$, $y = (a^-)^{-1} z$. (Similarly if $a = y^{-1} x$ with $x \geq e$, $y \geq e$, then $x = za^+$, $y = z(a^-)^{-1}$ with $z \geq e$.) Indeed, $z = x \wedge y$ satisfies $z \geq e$, and $a^- = a \wedge e = xy^{-1} \wedge e = (x \wedge y)y^{-1} = zy^{-1}$, whence $y = (a^-)^{-1}z$, and then $x =$

$= ay = (a^+ a^-) (a^-)^{-1} z = a^+ z$ (and similarly for $a = y^{-1} x$). If, in particular, $x \perp y$, then $x = a^+$, $y = (a^-)^{-1}$.

Lemma. *A l. o. group G is f. o. if, and only if, $a > e$ and $b > e$ imply $a \wedge b > e$.*

The "only if" part being clear assume the stated condition. Since $c^+ \wedge (c^-)^{-1} = e$ for every $c \in G$, we must have either $c^+ = e$ or $c^- = e$, establishing linearity.

G) $|a| \geq e$ *for all $a \in G$; equality holds if, and only if, $a = e$.* On multiplying the inequalities $|a| \geq a, |a| \geq a^{-1}$ we get $|a|^2 \geq e$ whence the desired inequality follows. If $|a| = e$, then $e \geq a$ and $e \geq a^{-1}$ whence $a = e$.

H) $|a^{-1}| = |a|$ *for all $a \in G$.*

I) $|ab| \leq |a| |b| |a|$, since

$$|a|^{-1}|b|^{-1}|a|^{-1} \leq |a|^{-1}|b|^{-1} \leq ab \leq$$
$$\leq |a| |b| \leq |a| |b| |a|.$$

J) *If G is commutative, then $|ab| \leq |a| |b|$.* In fact,

$$|ab| = ab \vee b^{-1} a^{-1} \leq ab \vee a^{-1} b \vee ab^{-1} \vee a^{-1} b^{-1} =$$
$$= (a \vee a^{-1})(b \vee b^{-1}) = |a| |b|.$$

This inequality is characteristic for commutative l. o. groups:[10] *if $|ab| \leq |a| |b|$ for all $a, b \in G$, then G is commutative.* Let $a, b \geq e$, then $ab = |ab| = |b^{-1} a^{-1}| \leq |b^{-1}| |a^{-1}| = ba$, and dually. The positive elements generate G, thus G is commutative.

K) $|a| = a^+ (a^-)^{-1}$ [or, by B), $|a| = a^+ (a^{-1})^+$]. This follows from

$$(a \vee e)(a^{-1} \vee e) = e \vee a \vee a^{-1} \vee e = |a| \vee e = |a|.$$

L) $|a^n| = |a|^n$ *for $n \geq 0$.* Owing to the permutability of a^+ and a^-, we obtain from K) and D) the desired $|a|^n = (a^+)^n (a^-)^{-n} = (a^n)^+ ((a^n)^-)^{-1} = |a^n|$.

M)[11] $|a \vee b| \leq |a| \vee |b| \leq |a| |b|$, for

[10] This has been observed by BUSULINI [1].
[11] If $a \perp b$ is assumed, then $|a| |b| = |a| \vee |b|$.

$$\left| a \vee b \right| = (a \vee b) \vee (a \vee b)^{-1} = (a \vee b) \vee (a^{-1} \wedge b^{-1}) \leqq$$

$$\leqq a \vee b \vee a^{-1} \vee b^{-1} = \left| a \right| \vee \left| b \right|$$

and G) implies $\left| a \right| \leqq \left| a \right| \left| b \right|$, $\left| b \right| \leqq \left| a \right| \left| b \right|$.

N) $\left| ab^{-1} \right| = (a \vee b)(a \wedge b)^{-1}$. This is a consequence of

$$(a \vee b)(a^{-1} \vee b^{-1}) = e \vee ab^{-1} \vee ba^{-1} \vee e = \left| ab^{-1} \right|.$$

O)[12] $\left| (a \vee c)(b \vee c)^{-1} \right| \left| (a \wedge c)(b \wedge c)^{-1} \right| = \left| ab^{-1} \right|$. By N)

and the distributivity, the left member is equal to

$$[(a \vee b) \vee c][(a \wedge b) \vee c]^{-1} [(a \vee b) \wedge c][(a \wedge b) \wedge c]^{-1}.$$

This may be written as $(tx \vee c)(x \vee c)^{-1}(tx \wedge c)(x \wedge c)^{-1}$
if we introduce the notations $x = a \wedge b$, $tx = a \vee b$. We have
to prove that the last expression is equal to

$$\left| ab^{-1} \right| = (a \vee b)(a \wedge b)^{-1} = t,$$

or equivalently, that

$$(x \vee c)^{-1}(tx \wedge c) = (tx \vee c)^{-1} t(x \wedge c)$$

holds. But

$$(x^{-1} \wedge c^{-1})(tx \wedge c) = x^{-1} tx \wedge x^{-1} c \wedge c^{-1} tx \wedge e$$

and

$$(x^{-1} t^{-1} \wedge c^{-1}) t(x \wedge c) = e \wedge x^{-1} c \wedge c^{-1} tx \wedge c^{-1} tc$$

are indeed equal, since because of $t \geqq e$, the terms $x^{-1} tx$ and
$c^{-1} tc$ may be omitted.

A large part of the above results can be generalized to certain
directed groups (FUCHS [1]). Here we restrict ourselves to the case
when G is a directed group whose order is isolated and, in addition,
distributive in the sense that

$$U(x_1, \ldots, x_m) U(y_1, \ldots, y_n) = U(x_1 y_1, x_1 y_2, \ldots, x_m y_n)$$

holds for all $x_i, y_j \in G$. We define a^+, a^- and $\left| a \right|$ as subsets of G:

$$a^+ = U(a, e), \qquad a^- = L(a, e), \qquad \left| a \right| = U(a, a^{-1}).$$

(Note that if G is a l. o. group, then the old and new concepts are
related as, in ring theory, elements and principal ideals are.) Let us
mention the following properties:

[12] See KALMAN [1].

1. $a^+ \subseteq P$, $a^- \subseteq P^{-1}$, $|a| \subseteq P$. Equality holds if, and only if, $a \leq e$, $a \geq e$ and $a = e$, respectively.

2. $(a^{-1})^+ = (a^-)^{-1}$, $(a^{-1})^- = (a^+)^{-1}$, $|a^{-1}| = |a|$.

3. $(ab)^+ \supseteq a^+ b^+$ and $(ab)^- \supseteq a^- b^-$.

4. $|ab| \supseteq |a| \, |b|$ for all $a, b \in G$ if, and only if, G is commutative.

5. $(a^+)^n_. = (a^n)^+$, $(a^-)^n = (a^n)^-$, $|a|^n = |a^n|$ for all natural integers n.

6. a^+ and $(a^-)^{-1}$ are permutable.

7. $|a| = a^+ (a^-)^{-1}$.

8. $|ab^{-1}| = U(a, b) L(a, b)^{-1}$.

The proofs are almost identical with those given for l. o. groups.

If positivity of complex numbers $z = x + yi$ is defined to mean $x > 0$ and $y \geq 0$, then a directed group is obtained with isolated and distributive order which is not a l. o. group.

5. *l*-ideals

An *l-ideal*[13] is defined to be an *o*-ideal in a l. o. group G; in other words, it is a normal and convex subgroup which is a sublattice of G. Hence an *l*-ideal A contains together with $a \in G$ also the elements a^+, a^- and $|a|$.

Theorem 8. *A is an l-ideal of a l. o. group G if, and only if, it is a normal subgroup of G such that*

$$a \in A, \quad x \in G \text{ and } |x| \leq |a| \quad imply \quad x \in A.$$

If a belongs to an *l*-ideal A and $x \in G$ satisfies $|x| \leq |a|$, then $-|a| \leq x \leq |a|$ together with $|a| \in A$ implies $x \in A$. Let, conversely, A be a normal subgroup with the stated property. If $e \leq x \leq a \in A$, then $|x| \leq |a|$ whence $x \in A$, and A is convex. If $a, b \in A$ are positive elements, then $e \leq \leq a \wedge b \leq a \vee b \leq ab \in A$ (see **4 M**), thus $a \wedge b$ and $a \vee b$ belong to A, i.e. the positive elements in A form a sublattice of A. Because of $e \leq a^+ \leq |a|$, $e \leq (a^-)^{-1} \leq |a|$, the positive elements in A generate A, and therefore Proposition 3 in Chapter II completes the proof.

A similar result can be proved for *o*-ideals in arbitrary p. o. groups: A is an *o*-ideal of G if, and only if, it is a directed normal subgroup of G and satisfies: $a \in A$, $x \in G$ and $|x| \supseteq |a|$ imply $x \in A$. Here $|a| = = U(a, a^{-1})$. (FUCHS [3].)

Note that every finitely generated *l*-ideal of a l. o. group G must be *principal*, i. e. generated by a single element. For the

---

[13] *l*-ideals have been used first by KANTOROVITCH [1]. See also RIESZ [1]. The general definition was given by BIRKHOFF [1] in terms of the property stated in Theorem 8.

l-ideal generated by the elements $g_1, \ldots, g_n$ contains $g = = \big| g_1 \big| \vee \ldots \vee \big| g_n \big|$, while the l-ideal generated by this g must contain all of g_i as it follows from $\big| g_i \big| \leq \big| g \big|$ and Theorem 8.

The principal l-ideal $I(a)$ generated by a can easily be described. Evidently, it contains every $x \in G$ satisfying $\big| x \big| \leq \leq \big| a_1 \big| \ldots \big| a_m \big|$ for some conjugates a_j of a. But the set I of all $x \in G$ with this property is an l-ideal containing a. In fact, $\big| xy \big| \leq \big| x \big| \big| y \big| \big| x \big|$ (see **4** I) implies that I is a subgroup which must be normal and which clearly fulfils the condition stated in Theorem 8. Thus

$$I(a) = [x \in G \big| \big| x \big| \leq \big| a_1 \big| \ldots \big| a_m \big| \text{ for conjugates } a_j \text{ of } a].$$

The importance of l-ideals is apparent from the following result.

Theorem 9. (Birkhoff [1].) *The kernel of a union (or intersection) preserving group homomorphism η of a l. o. group G onto a p. o. group H is an l-ideal A of G. The factor group of a l. o. group with respect to an l-ideal is again a l. o. group.*

If η preserves unions, then it preserves ordering and intersection too. Hence the kernel A of η is, by Theorem 7 of Chapter II, a normal, convex subgroup, and it is also a sublattice. If A is an l-ideal of G and $a \equiv b \pmod{A}$, then $\big| (a \vee c)(b \vee c)^{-1} \big| \leq \leq \big| ab^{-1} \big|$, $\big| (a \wedge c)(b \wedge c)^{-1} \big| \leq \big| ab^{-1} \big|$ (see **4** O) together with $ab^{-1} \in A$ imply that $a \vee c \equiv b \vee c$ and $a \wedge c \equiv b \wedge c \pmod{A}$. Hence the cosets of G mod A have the substitution property for both lattice operations, and so the natural map from G onto G/A preserves the order relation and lattice operations. Q. E. D.

Thus *the congruence relations of a l. o. group are the partitions into the cosets of its different l-ideals.*

Let us note that a group homomorphism η of a l. o. group G into a l.o. group is an o-homomorphism if, and only if, one of the following equivalent conditions is satisfied for all a, b in G: $\eta(a \vee b) = \eta(a) \vee \eta(b)$; $\eta(a \wedge b) = \eta(a) \wedge \eta(b)$; $\eta(\big| a \big|) = \big| \eta(a) \big|$; $\eta(a^+) = \eta(a)^+$; $a \perp b$ implies $\eta(a) \perp \eta(b)$ (Birkhoff—Pierce [1]).

Next consider two l-ideals A and B of G. Obviously, $A \cap B$ is again an l-ideal. But it is not so evident that the product

AB (i. e. the set of elements ab with $a \in A$, $b \in B$) is again an l-ideal of G. In order to prove this, it suffices to verify that an $x \in G$ satisfying $|x| \leq |ab|$ for some $a \in A$, $b \in B$ belongs to AB. From 4 I) we obtain $|x| \leq |a| \, |b| \, |a|$ whence, by Corollary 2, there exist positive elements $a_1, a_2 \leq |a|$ and $b_1 \leq |b|$ such that $|x| = a_1 b_1 a_2$. Here, by convexity, $a_1, a_2 \in A$, $b_1 \in B$, whence $|x| \in AB$. Because of $e \leq x^+ \leq |x|$, a similar reasoning applies to show that also $x^+ \in AB$, whence $x \in AB$ as well. We obtain

Lemma A. (BIRKHOFF [1].) *The subgroup of a l. o. group G generated by l-ideals of G is again an l-ideal.*

Now it is easy to derive:

Theorem 10. (BIRKHOFF [1].) *The l-ideals of a l. o. group G form a complete and distributive sublattice in the lattice of all normal subgroups of G.*[14]

What we have to prove follows from complete distributivity:

$$(1) \qquad A \cap \{\ldots, B_\lambda, \ldots\} = \{\ldots, A \cap B_\lambda, \ldots\}$$

for l-ideals $A, \ldots, B_\lambda, \ldots$ of G. Here only the inclusion $\subseteq$ needs a verification. Assume that $a > e$ belongs to the left member, that is, $a = b_1 \ldots b_m$ with $a \in A$, $b_i \in B_i$. Because of $b_k b_{k+1} \ldots b_m \equiv b_{k+1} \ldots b_m \pmod{B_k}$ we have $x_{k-1} \equiv x_k \pmod{B_k}$ if x_k is defined as $x_k = a \wedge (e \vee b_{k+1} \ldots b_m)$, $k = 0, 1, \ldots, m$. Observe that x_k belongs to A because of $e \leq x_k \leq a$, and so $x_{k-1} x_k^{-1} \in A \cap B_k$. Since $x_0 = a$ and $x_m = e$, the identity $a = (x_0 x_1^{-1})(x_1 x_2^{-1}) \ldots (x_{m-1} x_m^{-1})$ proves that a belongs to the right member of (1). The same must hold for every element of the left member, completing the proof.

From the last result it follows at once that if

$$G = G_1 \times \ldots \times G_n$$

is a direct decomposition[15] of a l. o. group G and A is an l-ideal of G, then

---

[14] A similar result has been proved by ŠIK [1] for the sets X^* defined in Proposition 12 and by LORENZ [1] for the subgroups $J(a) = [x \in G \mid |x| \leq |a|^n$ for some n].

[15] This direct decomposition is to be understood both in the group- and lattice-theoretical senses.

$$A = A_1 \times \ldots \times A_n \quad \text{with} \quad A_i = G_i \cap A$$

is a direct decomposition of A.

Let us formulate the following simple observation:

Lemma B. *Let G be a l. o. group and $A_1, \ldots, A_n$ l-ideals of G. If G as abstract group is the direct product of the A_i, then G as l. o. group is also the direct product of the l. o. groups A_i.*

What we have to show is that $a_1 a_2 \ldots a_n \geq e$ $(a_i \in A$) only if $a_i \geq e$ for all i. Hypothesis implies

$$a_1^+ \ldots a_{i-1}^+ \, a_i^+ \, a_i^- a_{i+1}^+ \ldots a_n^+ \geq e,$$

i. e.

$$a_1^+ \ldots a_i^+ \ldots a_n^+ \wedge (a_i^-)^{-1} = (a_i^-)^{-1}.$$

But $a_j^+ \wedge (a_i^-)^{-1} \in A_j \cap A_i = e$ $(j \neq i)$, so that from **2** A) we get $a_i^+ \wedge (a_i^-)^{-1} = (a_i^-)^{-1}$. Hence **4** F) implies $a_i^- = e$, i. e. $a_i \geq e$.

The positive cone $P(A) = P \cap A$ of an l-ideal A of G is obviously a normal and convex subsemigroup of the positive cone P of G such that $e \in P(A)$. We have

Theorem 11. (CONRAD [12].) *The correspondence*

$$A \to P(A)$$

is one-to-one between all l-ideals A of G and all normal and convex subsemigroups S of P containing e. The inverse correspondence is

$$S \to \{S\}.$$

Evidently $\{P(A)\} = A$. Let S have the stated properties, and $A = \{S\}$. Every $a \in A$ has the form $a = bc^{-1}$ with $b, c \in S$, where we may assume that $b \perp c$, for b and c can be replaced by $b(b \wedge c)^{-1}$ and $c(b \wedge c)^{-1}$, respectively, owing to the convexity of S. Then $a^+ = b$; and so $a \in P(A)$ if, and only if, $a = a^+ = b$. Thus $P(A) = S$, and the convexity of A also results. $P(A)$ is a sublattice, for S is one: $e \leq b \wedge c \leq b \vee c \leq bc$ for all $b, c \in S$.

We close this section with the following result. Call the arbitrary elements a, b of G orthogonal if $|a|$ and $|b|$ are orthogonal in the previous sense, i. e.

$$|a| \wedge |b| = e.$$

It will always be clear from the context whether we refer to orthogonality in this general sense or in the old one.

Proposition 12.[16] *In a l. o. group G, the set X^* of all elements orthogonal to every element of a subset X is a convex subgroup and a sublattice. X^* is an l-ideal for every subset X of G if, and only if, the carriers of G are invariant.*

Let $b, c \in X^*$. Because of $|bc^{-1}| \leq |b| |c^{-1}| |b| \perp |a|$ for all $a \in X$, the element bc^{-1} also belongs to X^*. Thus X^* is a subgroup. By 4 O) we have $|(b \vee c)c^{-1}| |(b \wedge c)c^{-1}| = |bc^{-1}|$ whence both $|(b \vee c)c^{-1}|$ and $|(b \wedge c)c^{-1}|$ are orthogonal to X, and so $b \vee c$, $b \wedge c \in X^*$. The convexity of X^* is evident.

The "if" part of the second assertion is evident, while the converse follows immediately by taking for X a one-element subset.

6. Groups with a finite number of carriers

Now we investigate the special case in which the l. o. group G has but a finite number of carriers. We shall see that in this case G can be obtained from f. o. groups by successive applications of direct products and lexicographic extensions.[17]

We begin with the rather elementary

Proposition 13. (JAFFARD [6].) *If the lattice $\mathfrak{C}$ of carriers of a l. o. group G is finite, then it is a Boolean algebra.*

From Theorem 5 we know that $\mathfrak{C}$ is a distributive lattice. Let $\mathfrak{C}$ contain n atoms $a_1^\wedge, \ldots, a_n^\wedge$ $(a_i \in G)$. If $b^\wedge \in \mathfrak{C}$ and say $a_1^\wedge, \ldots, a_k^\wedge \leq b^\wedge$, but $a_{k+1}^\wedge, \ldots, a_n^\wedge \nleq b^\wedge$, then $c^\wedge = a_{k+1}^\wedge \vee \ldots \vee a_n^\wedge$ is the complement of $b^\wedge$ in $\mathfrak{C}$. For, on the one hand, $b^\wedge \wedge c^\wedge = (b^\wedge \wedge a_{k+1}^\wedge) \vee \ldots \vee (b^\wedge \wedge a_n^\wedge) = e^\wedge$. On the other hand $u = b \vee c$ $(b \in b^\wedge, c \in c^\wedge)$ satisfies $a_i^\wedge \leq b^\wedge \vee c^\wedge = (b \vee c)^\wedge = u^\wedge$ for all i whence $u \wedge x = e$ implies $a_i \wedge x = e$ for all i, i. e. x contains no atoms and so $x = e$, $u^\wedge$ is the maximal element of $\mathfrak{C}$. Consequently, $\mathfrak{C}$ is a Boolean algebra, in fact.

---

[16] The first part is due to BIRKHOFF [1].
[17] See BIRKHOFF [1], JAFFARD [7] (only commutative groups), CONRAD—CLIFFORD [1] and CONRAD [12].

The number n of atoms in $\mathfrak{C}$ can be characterized as the maximal number n such that G contains n pairwise orthogonal elements $\neq e$.

For a subset $N = [i_1, \ldots, i_k]$ of $1, 2, \ldots, n$ we define $G_N = G_{i_1 \ldots i_k}$ and $G_N^* = G_{i_1 \ldots i_k}^*$ as the subgroups generated by elements belonging to carriers $\leq a_{i_1}^\frown \vee \ldots \vee a_{i_k}^\frown = a_N^\frown$ and $< a_N^\frown$, respectively.

Lemma A. (JAFFARD [7], CONRAD [12].) *For the subgroups G_N, G_N^* the following assertions are true*:

(i) *they are convex subgroups and sublattices of G*;

(ii) *G_N^* is an l-ideal of G_N*;

(iii) *if $G_N^* \neq e$, then G_N is a lexicographic extension of G_N^* by the f. o. group G_N/G_N^*.*

In order to verify (i), let $e \leq x \leq b_1^{\pm 1} \ldots b_r^{\pm 1} \leq b_1 \ldots b_r$ with $b_i^\frown \leq a_N^\frown$ (or $b_i^\frown < a_N^\frown$). From Corollary 2 we know that $x = c_1 \ldots c_r$ for some $e \leq c_i \leq b_i$ whence $c_i^\frown \leq a_N^\frown$ (or $c_i^\frown < a_N^\frown$). Therefore x belongs to G_N (or to G_N^*). Now if $y = d_1^{\pm 1} \ldots d_s^{\pm 1}$, $z = d_{s+1}^{\pm 1} \ldots d_t^{\pm 1}$ with $d_i^\frown \leq a_N^\frown$ (or $d_i^\frown < a_N^\frown$), then y, z, and hence $y \wedge z, y \vee z$ lie in the interval $[d_1^{-1} \ldots d_t^{-1}, d_1 \ldots d_t]$. Convexity establishes (i).

It suffices to verify (ii) and (iii) in the case $G_N = G$ and $G_N^* = G^*$ is a non-trivial subgroup of G. G^* is a normal subgroup, for the inner automorphisms of G permute the non-maximal carriers among themselves. By (i) this implies (ii).

Evidently, every $g \in G \setminus G^*$ $(g > e)$ belongs to the maximal carrier. If $a \in G^*$ $(a > e)$ and if we write $g = (g \wedge a)g_1$, $a = (g \wedge a)a_1$, then necessarily $g_1 \wedge a_1 = e$. Since $e \leq g \wedge a \leq a$ and so $g \wedge a \in G^*$, we have $g_1 \notin G^*$ whence $g_1^\frown = g^\frown$. Therefore $g_1 \wedge a_1 = e$ implies $a_1 = e$, that is, $g > a$ for all $a \in G^*$, and G is a lexicographic extension of G^*. Finally, if G/G^* were not f. o., then there would exist positive elements x, $y \notin G^*$ such that $x \wedge y \in G^*$. Since we have assumed $G^* \neq e$, there exists a $z \in G^*$ satisfying $z > x \wedge y$. But by what has been shown it follows that $z < x$, $z < y$, i. e. $z \leq x \wedge y$, a contradiction. This completes the proof.

Lemma B. (JAFFARD [7], CONRAD [12].) *The subgroups $G_1, \ldots, G_n$ are f. o. convex subgroups of G and the subgroup S*

they generate is an l-ideal of G such that

$$S = G_1 \times \ldots \times G_n.$$

From (i) of Lemma A we conclude that the G_i are l. o. convex subgroups of G. The proof of (i) shows that the positive cone of G_i is the semigroup (with e) generated by the elements in $a_i^{\wedge}$. Now if $x, y \in a_i^{\wedge}$ (i. e. $x^{\wedge} = y^{\wedge} = a_i^{\wedge}$), then $x \wedge y = e$ is impossible, since this would imply $a_i \wedge y = e$, $a_i \wedge a_i = e$. Thus each G_i is f. o. The inner automorphisms of G carry the atoms of $\mathfrak{C}$ into atoms, therefore $S = \{G_1, \ldots, G_n\}$ is a normal subgroup of G. As in the proof of (i) of the preceding lemma it follows that S is convex, and by 4 M) we get that it is a sublattice as well. The positive elements of G_i are orthogonal to those of G_j $(i \neq j)$, for $(a_i \wedge a_j)^{\wedge} = a_i^{\wedge} \wedge a_j^{\wedge} = e^{\wedge}$. Consequently, S as abstract group is the direct product of the G_i, and so Lemma B in **5** completes the proof.

Lemma C. (CONRAD [12].) *The lattice of carriers of the l. o. factor group G/S contains fewer than n atoms.*

Let $b \in G \setminus S$ $(b > e)$ and let $b^{\wedge} = a_N^{\wedge}$ for some $N = [i_1, \ldots, i_m]$. Then either $b \in G_N \setminus G_N^*$ and $m \geq 2$, or $b \in G_N^* = \{\ldots, G_{N'}, \ldots\}$ where N' runs over all subsets of N consisting of $m - 1$ elements and $m \geq 3$. In either case $b > a$ for all $a \in G_i \times G_j$ for some indices $i \neq j$. In the first alternative this follows immediately from Lemma A, (iii), while in the second case an obvious induction is necessary.—Now if we assume, by way of contradiction, that $G \setminus S$ contains n elements $b_1, \ldots, b_n$ $(> e)$ such that $b_k \wedge b_l \in S$ for all $k \neq l$, then from what has been said of b we infer that there exist three indices r, s, t $(r \neq s)$ such that both b_r and b_s are greater than every element of G_t. But $b_r \wedge b_s \in S$ cannot be greater than every element of G_t.

Lemma D. (CONRAD [12].) *For each subset N of $1, 2, \ldots, n$ containing at least two elements, there exists a proper direct decomposition*

(1) $$G_N^* = G_{N_1} \times G_{N_2}$$

with $N_1 \cup N_2 = N$.

Again, it suffices to consider the case $G_N^* = G^*$. We use induction on n. For $n = 2$ we have $S = G_1 \times G_2 = G^*$; assume $n > 2$ and the assertion true for l. o. groups with fewer than n atoms in the lattice of carriers. By Lemma C and induction hypothesis we have

$$(G^*/S)^* = U/S \times V/S$$

where either U/S is f. o. and $V = S$ or U, V properly contain S.

Assume that the group $(G^*/S)^*$ is properly contained in G^*/S. Then the corresponding factor group G^{**} is, because of Lemma A, (iii), f. o. Let us consider the images of $G_{23\ldots n}, \ldots,$ $G_{12\ldots n-1}$ under the natural mapping φ of G^* onto G^{**}. They are convex subgroups and generate G^{**}, thus one of the images coincides with G^{**}. If $\varphi(G_{23\ldots n}) = G^{**}$, then $G^* = G_1 \times G_{23\ldots n}$, since here the factors generate G^*, are convex sublattices of G, further because of orthogonality they are disjoint and commute (cf. also Lemma B of 5). Henceforth we may restrict ourselves to the case $G^*/S = U/S \times V/S$.

The same reasoning establishes the case in which G^*/S is f. o.

If neither $U = S$ nor $V = S$, then define N_u as the set of all indices i for which a $u \in U \setminus S$ exists with the property that $u > a$ for all $a \in G_i$, and let N_u' be the set of all other indices. G_{N_u} and

$$H_u = \mathop{\textit{II}}_{j \in N_u'} G_j$$

together generate U. For if $u \in U$ $(u > e)$ and $u \in G_M \setminus G_M^*$ for some M containing at least two indices, then clearly $M \subseteq N_u$, and so $u \in G_{N_u}$. If $u \in U$ $(u > e)$, but u does not belong to any such $G_M \setminus G_M^*$, then it is the product of positive elements belonging to some $G_M \setminus G_M^*$ or to some G_l, and therefore $u \in \{G_{N_u}, H_u\}$. The sets N_u, N_u' are clearly invariant under inner automorphisms of G^*; thus G_{N_u} and H_u are normal subgroups of G^*. They are obviously convex sublattices of G, i. e. they are l-ideals of G^*. Finally, they are disjoint, since

$$G_{N_u} \cap H_u \subseteq G_{N_u} \cap S \cap H_u = \mathop{\textit{II}}_{i \in N_u} G_i \cap H_u = e.$$

Therefore, from Lemma B in **5** we conclude that $U = G_{N_u} \times \times H_u$. — Next define N_v and N'_v similarly for V; then $V = = G_{N_v} \times H_v$. If $i \in N_u \cap N_v$, then there exist $u \in U \setminus S$ and $v \in V \setminus S$ such that $u > a$, $v > a$ for all $a \in G_i$. Thus $u \wedge v \in \in U \cap V = S$, but $u \wedge v > a$ for all $a \in G_i$ is impossible in S. Hence N_u and N_v are disjoint. We prove that

$$(2) \qquad G^* = G_{N_u} \times G_{N_v} \times H \quad \text{with} \quad H = H_u \cap H_v.$$

Here the components are clearly l-ideals of G^* which together generate G^*. It is also obvious that G^* as abstract group is their direct product, hence Lemma B of **5** implies (2). Now if we write $N_1 = N_u$, $N_2 = N'_u$, then in view of Theorem 10 and

$$G_{N_2} = G^* \cap G_{N_2} = (G_{N_1} \cap G_{N_2}) \times G_{N_v} \times H = G_{N_v} \times H$$

we obtain (1).

Now we are ready to prove the main result:

Theorem 14. (CONRAD [12].) *If G is a l. o. group with a finite number of carriers, then G can be obtained from a finite number of f. o. groups by an alternating sequence of direct products and lexicographic extensions where the factor groups in the lexicographic extensions are always f. o.*

Proof by induction on the number n of atoms in the lattice of carriers of the l. o. group G. If $n = 1$, then G is, by Lemma B, f. o., and we are ready. If $n \geq 2$, then either $G = G^*$ or G is, by Lemma A, a lexicographic extension of G^* by a f. o. group G/G^*. Here G^* is, by Lemma D, a direct product of two l. o. groups with fewer than n atoms in the lattices of carriers. Q. E. D.

For instance, if $n = 2$, then G is either the direct product of two f. o. groups or the lexicographic extension of such a direct product by a f. o. group. If $n = 3$, then G is either the direct product of a l. o. group with $n = 2$ and a f. o. group or the lexicographic extension of such a group by a f. o. group. For explicit examples we refer to CONRAD [12].

Theorem 14 can be generalized to l. o. groups in which every carrier is greater than at most a finite number of carriers. Cf. CONRAD [14].

7. Units

A *weak unit* of a l. o. group G is an element $>e$ which is orthogonal only to e. A *strong unit* is an element $u \in G$ such that to each $a \in G$ there exists a natural integer n satisfying $u^n > a$.[18]

In the l. o. (additive) group of all continuous real functions defined in $[0, \infty)$, the function $f(x) \equiv 1$ is a weak unit, but there exists no strong unit. $f(x) \equiv 1$ is a strong unit in the l. o. group of all bounded real functions.

A) *In a f. o. group every element $>e$ is a weak unit.*

B) *A l. o. group contains a weak unit if, and only if, the lattice of its carriers has a maximal element.*

C) *Every strong unit is a weak unit.* If u is a strong unit, then $u^n > e$ for some positive integer n whence, by the isolated character of lattice order, we obtain $u > e$. Assume that $u \perp a$. Then $u^m \perp a$ for every natural m. If m is chosen such that $u^m > a$, then we get $a = e$, as desired.

D) *A f. o. group has a strong unit if, and only if, it has a maximal convex subgroup.* If u is a strong unit in the f. o. group G, then $\{u\}_{\square} = G$ and the union C of all convex subgroups not containing u is the maximal convex subgroup of G. Conversely, if G has a maximal convex subgroup C, then this C must be normal and G/C is Archimedean. Every positive element in $G \setminus C$ is then a strong unit of G.

E) *If the elements of the l. o. group G which belong to non-maximal carriers generate a proper subgroup G^* of G and the f. o. factor group G/G^* has a strong unit, then G has one too.* Lemma A, (iii) in **6** holds for arbitrary l. o. groups, hence our group G is the lexicographic extension of G^* by the f. o. group G/G^*. If uG^* $(u > e)$ is a strong unit of G/G^*, then u is a strong unit of G, since $u > a$ holds for all $a \in G^*$.

F) *If the l-ideal I of G has a strong unit v, then I is the principal l-ideal $I(v)$ generated by v.* In fact, $I(v)$ contains every positive element a of I, since we have $e \leq a < v^n$ for some n.

---

[18] For these concepts see FREUDENTHAL [1] and BIRKHOFF [1]. Weak units play an important role in applications where integral representations may be given in terms of a certain "resolution" of a weak unit. See e. g. BIRKHOFF [3].

G) *In a commutative l. o. group, the element* $|a|$ *is a strong unit for the l-ideal* $I(a)$ *generated by* a. Since $I(a)$ consists of all $x \in G$ such that $|x| \leq |a|^m$ for some $m > 0$ (see **5**), it is obvious that $|a|$ is a strong unit of $I(a)$.

This statement does not hold, in general, in the non-commutative case, as shown by JAKUBÍK [1]. He also established the existence of a l. o. group, generated by two elements in the group-theoretic sense, which has only one non-trivial *l*-ideal and this contains no strong unit.

H) *In a l.o. group, the l-ideals having strong units and* $\{e\}$ *form a sublattice of the lattice of all l-ideals*. It is enough to verify that if the *l*-ideals I_1, I_2 have strong units v_1, v_2, then both $I = I_1 \cap I_2 \neq \{e\}$ and $I^* = \{I_1, I_2\}$ have strong units. If $a \in I$, then $a < v_1^m$ and $a < v_2^n$ for certain integers m, n. Since $v_1, v_2 > e$, we have $v_1^m \wedge v_2^n < (v_1 \wedge v_2)^{m+n}$ and therefore $v_1 \wedge v_2$ is a strong unit of I. If $bc \in I^*$ where $b \in I_1$, $c \in I_2$, then $b < v_1^m$ and $c < v_2^n$ for some integers m, n. Hence $bc < v_1^m v_2^n \leq (v_1 \vee v_2)^{m+n}$, thus $v_1 \vee v_2$ is a strong unit of I^*.

8. Lattice-ordered vector groups

The complete direct product

$$G = \prod_{\lambda \in \Delta}^{*} G_{\lambda}$$

of f. o. groups G_{λ} is a l. o. group. A subgroup of such a G will be called a *l. o. vector group* if it is a sublattice of G as well. Observe that a l. o. group which is a vector group in the sense of Chapter III, **6** need not be a l. o. vector group. Criteria of a rather simple character can be given for deciding whether a l. o. group is a l. o. vector group.

Theorem 15.[19] *For a l. o. group* G *the following conditions are equivalent:*

(a) *if* $a \in G$ *is orthogonal to one of its conjugates* $x^{-1} ax$, *then* $a = e$;

---

[19] LORENZEN [3] established the equivalence of (a) and (c); he called l. o. groups satisfying (a) regular. For a simpler proof see LORENZ [1]. ŠIK [4] proved that (b) and (c) are equivalent.

(b) *the carriers are invariant*;

(c) *G is a l. o. vector group.*

First of all we show that (c) implies (a). If G is a l. o. vector group, then the orthogonality of a and $x^{-1} ax$ is equivalent to the orthogonality of their λth components for all λ. But in a f. o. group (a) is trivially true, therefore every component of a must be the identity, that is, $a = e$.

By virtue of Proposition 6, it suffices to verify that (b) implies (c). By Proposition 7, every l. o. group G with invariant carriers is (group- and lattice-theoretically isomorphic to) a subdirect union of subdirectly irreducible l. o. groups G_λ with invariant carriers. Assume that some G_λ is not f. o., i. e. $b > e$ and $c > e$ exist with $b \wedge c = e$. The set B of all elements of G orthogonal to c and the set C of all elements orthogonal to B are, by Proposition 12, l-ideals of G_λ. Clearly, $b \in B$, $c \in C$ and $B \cap C = e$, in contradiction to the subdirect irreducibility of G_λ. Hence all the G_λ are f. o. and G is a l. o. vector group.

We obtain immediately: *a commutative l. o. group is a l. o. vector group* (CLIFFORD [1]).

JAFFARD [6] seeks conditions for the irredundancy of a subdirect product representation of a l. o. vector group. He finds a necessary and sufficient condition, in terms of the carriers, for the existence of such an irredundant representation. Elsewhere [8] he investigates groups which are complete sublattices of some $\prod^* G_\lambda$ with f. o. G_λ.

ŠIK [1], [2] gives necessary and sufficient conditions for a l. o. group to be a direct product (or a complete direct product) of f. o. groups. For special representations of l. o. vector groups see also ŠIK [4].

JAKUBÍK [5] has shown that the possibility of representing a l. o. group G as a complete direct product of f. o. groups can already be decided merely in terms of lattice-theoretical properties of the positive cone of G, but the same does not hold for the subdirect products.

9. Complete lattice-ordered groups

A p. o. group G is said to be (conditionally) *complete*[20] if every non-void set in G bounded from above has a l. u. b. in G. An obviously equivalent condition is that the non-void subsets

[20] We shall omit the adverb "conditionally", since a non-trivial l. o. group cannot be complete as a lattice.

bounded from below have a g. l. b. in G. The complete l. o. groups are of great importance in analysis.

Let us consider some algebraic rules in complete l. o. groups. If $\vee x_a$ (if $\wedge x_a$) exists,[21] then

$$(1) \qquad\qquad a(\vee x_a)b = \vee \, ax_a \, b,$$

$$(2) \qquad\qquad a(\wedge x_a)b = \wedge \, ax_a \, b,$$

$$(3) \qquad\qquad (\vee x_a)^{-1} = \wedge \, x_a^{-1}$$

for all $a, b, x_a \in G$; this follows from Chapter II, **1**, (ii) and (iv).

We verify the infinite distributive rules:[22]

$$(4) \qquad\qquad a \wedge (\vee x_a) = \vee (a \wedge x_a),$$

$$(5) \qquad\qquad a \vee (\wedge x_a) = \wedge (a \vee x_a)$$

for all $a, x_a \in G$. It suffices to prove (4); then (5) follows by duality. Putting $v = \vee x_a$, we have $e \leq (a \wedge v)\,(a \wedge x_a)^{-1} \leq \; \leq vx_a^{-1}$ by **4** O). Since $\wedge(vx_a^{-1}) = v(\vee x_a)^{-1} = e$, we have

$$e = \wedge((a \wedge v)\,(a \wedge x_a)^{-1}) = (a \wedge v)\,(\vee(a \wedge x_a))^{-1},$$

whence (4) follows.

Lemma. (KANTOROVITCH [1].) *A complete p. o. group is completely integrally closed.*

Assume that $a^n \leq b$ for $n = 1, 2, \ldots$. By completeness, the element $c = a \vee a^2 \vee \ldots \vee a^n \vee \ldots$ exists in G, and we have

$$ac = a \cdot \bigvee_{n=1}^{\infty} a^n = \bigvee_{n=2}^{\infty} a^n \leq c.$$

Hence $ac \leq c$, and $a \leq e$, as required.

A subset H of a complete l. o. group G will be called *closed*[23] if $x_a \in H$ and the existence of $\vee x_a$ in G imply $\vee x_a \in H$. If H is a subgroup, the same holds for $\wedge$. Thus a closed subgroup is at the same time a sublattice.

---

[21] Here $\vee$ and $\wedge$ denote complete joins and meets, respectively.
[22] See KANTOROVITCH [1].
[23] RIESZ [1].

From (4) it follows that in complete l. o. groups, the sets X^* as defined in **5** are closed. Moreover, they are direct factors:

Theorem 16. (RIESZ [1], BIRKHOFF [3].) *In a complete l. o. group G, a convex, closed subgroup J is a direct factor of G. Specifically,*

$$G = J \times J^*$$

where J^* consists of all elements of G which are orthogonal to J.

We show that the direct decomposition $G = J \times J^*$ holds. By Proposition 12, J^* is a convex subgroup and a sublattice, and evidently, $J \cap J^* = e$. Since the positive elements of J and J^* generate J and J^*, respectively, and as positive orthogonal elements are permutable, we infer that J and J^* are element-wise permutable. For a positive $a \in G$ define $a_1 = = \bigvee (a \wedge t_a)$ with t_a running over all elements of J. By hypothesis, $a \wedge t_a \in J$ implies $a_1 \in J$. Define $a_2 = a_1^{-1} a$ which is clearly $\geq e$. For arbitrary $t \in J$, $|t| = t \vee t^{-1} \in J$, $a_1 |t| \in J$, therefore $a_1 \geq a \wedge (a_1 |t|) = a_1 (a_2 \wedge |t|)$, whence $a_2 \wedge |t| = e$ for all $t \in J$, i. e. $a_2 \in J^*$. We conclude that the positive cone of G is the direct product of the positive cones of J and J^*, consequently, $G = J \times J^*$ both group- and lattice-theoretically.

The intersection of a set of closed l-ideals is again closed. Therefore from Theorems 10 and 16 one obtains

Corollary 17. (BIRKHOFF [3].) *The closed l-ideals of a complete l. o. group form a Boolean algebra.*

Observe that in Theorem 16 the normality of J was a consequence of the hypotheses. This fact will be made use of in the proof of the next important result.

Theorem 18. (IWASAWA [1], OGASAWARA [1].) *A complete l. o. group G is commutative.*

The basic idea of the following proof (given by BIRKHOFF [3]) is essentially a modification of a proof of Theorem 1 in Chapter IV.

We begin with the observation that to every finite set $x_1, \ldots, x_n$ of elements of G we can find a direct decomposition $G = J_1 \times \ldots \times J_k$ such that all the components of $x_1, \ldots, x_n$ in the closed l-ideals J_i are comparable with e, for every i.

By Proposition 12 and Theorem 16, $(x_j^+)^*$ is a direct factor of G whose complement $(x_j^+)^{**}$ contains x_j^-. By the distributivity of the lattice of direct factors we may take a common refinement to obtain a decomposition of G, as required.

In order to establish commutativity, take two positive elements a, b of G. On account of the preceding observation we may suppose that $a^{-1}b$, $[a, b]$, $[a, b, b]$ are comparable with e. By symmetry, we may assume $a^{-1}b > e$. In case $[a, b] \geq e$ we distinguish two subcases according as $[a, b, b] \geq e$ or $\leq e$. In the first subcase $[a, b] \leq b^{-1}[a, b]b$, and so $[a, b] \leq b^{-i}[a, b]b^i$ for $i \geq 0$. Hence, by the identity

$$[a, b^n] = \prod_{i=0}^{n-1} (b^{-i}[a, b]b^i) \qquad (n > 0)$$

we infer that $[a, b]^n \leq [a, b^n]$ for $n = 1, 2, \ldots$. Since $a < b$ implies $b^{-n}ab^n < b$, we have $[a, b^n] < a^{-1}b$. From $[a, b]^n < a^{-1}b$ $(n = 1, 2, \ldots)$ we obtain $[a, b] \leq e$, using the lemma, and so $[a, b] = e$. In the second subcase $[a, b] \leq b^i[a, b]b^{-i}$ for $i \geq 0$, and from the identity

$$[b^{-n}, a] = \prod_{i=n}^{1} (b^i[a, b]b^{-i}) \qquad (n > 0)$$

we get $[a, b]^n \leq [b^{-n}, a]$ for $n = 1, 2, \ldots$. Hence $[a, b]^n \leq b^n a^{-1}b^{-n}a < a$ and complete integral closure imply $[a, b] \leq e$ and $[a, b] = e$. The remaining case $[a, b] \leq e$ can be disposed of similarly. We conclude that a and b commute.

It can be shown that a complete l. o. group is the direct product of two groups, one of which is o-isomorphic to a subdirect product of f. o. infinite cyclic groups and the other is a complete vector lattice. Here a *vector lattice* is a l. o. group which is a vector space over the real numbers such that multiplication by positive real numbers preserves order.

10. Embedding in complete lattice-ordered groups

We turn to the problem of embedding a l. o. group in a complete one.

The classical method of forming Dedekind cuts can be applied to obtain an embedding of an arbitrary p. o. set in a complete lattice—as shown by MacNeille.[24] If this process is applied to a p. o. group, the result is in general not a group. But if the p. o. group can be embedded at all in a complete l. o. group with preservation of order, as in the case of rationals, then the embedding can be achieved by the Dedekind process.

Let G be an arbitrary p. o. group. We shall consider only non-void, u-bounded subsets X of G. The correspondence

$$X \to X^\sharp = L(U(X))$$

is a closure operation, i. e. it has the properties:

$$X \subseteq X^\sharp; \quad X^{\sharp\sharp} = X^\sharp; \quad X \subseteq Y \text{ implies } X^\sharp \subseteq Y^\sharp.$$

The set $G^\sharp$ of all "closed" sets C (i. e. neither C nor $U(C)$ is void and $C^\sharp = C$) is a conditionally complete conditional lattice[25] under set-inclusion. In fact, (a) if $C_0 \subseteq C_a$ ($\in G^\sharp$), then the intersection $\cap C_a$ lies in $G^\sharp$ and is the g. l. b. $\wedge C_a$ for the C_a:

$$\cap C_a = \cap L(U(C_a)) = L(\cup U(C_a)) \supseteq L(U(\cap C_a)) \supseteq \cap C_a,$$

(b) if $C_0 \supseteq C_a$ ($\in G^\sharp$), then the closure of the set-union: $(\cup C_a)^\sharp$ is the l. u. b. $\vee C_a$ for the C_a. The correspondence

$$a \to a^\sharp = L(a)$$

embeds G in $G^\sharp$. It preserves order, g. l. b. and l. u. b., for if $\wedge a_a$ exists in G, then

$$(\wedge a_a)^\sharp = L(\wedge a_a) = L(\ldots, a_a, \ldots) = \cap L(a_a) = \wedge a_a^\sharp,$$

and if $\vee a_a$ exists in G, then $a_a^\sharp \leq C$ ($\in G^\sharp$) for all a implies $a_a \in C$, $\vee a_a \in C$, $(\vee a_a)^\sharp \leq C$, and so (because of $a_a^\sharp \leq \leq (\vee a_a)^\sharp$) $(\vee a_a)^\sharp$ is the l. u. b. for the $a_a^\sharp$ in $G^\sharp$:

$$(\vee a_a)^\sharp = \vee a_a^\sharp.$$

$G^\sharp$ is a lattice exactly if G is directed.

[24] *Trans. Amer. Math. Soc.*, **42** (1937), 416—460.
[25] By a conditional lattice we mean a p. o. set in which every two elements that have an upper (lower) bound have a l. u. b. (g. l. b.).

Define the product $X \cdot Y$ of $X, Y \in G^{\#}$ as the closure of the set XY of all xy $(x \in X,\ y \in Y)$.[26] One verifies easily that

$$a^{\#} \cdot b^{\#} = (ab)^{\#}, \quad a^{\#} \cdot C = (aC)^{\#}, \quad C \cdot a^{\#} = (Ca)^{\#},$$

and $e^{\#}$ is the neutral element of $G^{\#}$. Since

$$X^{\#} \cdot Y = (X^{\#} Y)^{\#} = (\bigcup_{y} X^{\#} y)^{\#} = (\bigcup_{y} (Xy)^{\#})^{\#} =$$

$$= \vee (Xy)^{\#} \subseteq (XY)^{\#}$$

for $X \subseteq G,\ Y \in G^{\#}$, and so

$$X^{\#} \cdot Y = (XY)^{\#},$$

we conclude that multiplication is associative in $G^{\#}$. Further $A \leq B$ implies $C \cdot A \leq C \cdot B$ and $A \cdot C \leq B \cdot C$ for all A, B, C in $G^{\#}$. We see that $G^{\#}$ is a semigroup and a conditional lattice[27] in which G is embedded as a subgroup-sublattice.

In general, $G^{\#}$ is not a group. In fact,

Lemma.[28] $C \in G^{\#}$ *has a left-inverse* D *in* $G^{\#}$ *if, and only if, one of the following equivalent conditions holds*:

(i) $U(L(C^{-1})C) = P$;

(ii) $L(C^{-1}U(C)) = P^{-1}$;

(iii) $U(C)x \subset U(C)$ *(if and) only if* $x \in P$.

Then $D = L(C^{-1})$.

From (i) we get $L(C^{-1}) \cdot C = L(U(L(C^{-1})C)) = L(P) = e^{\#}$, i. e. $D = L(C^{-1})$ is a left-inverse of C. Conversely, let C have a left-inverse D, $L(U(DC)) = P^{-1}$. From $Dc \subseteq P^{-1}$, i. e. from $D \subseteq P^{-1}c^{-1}$ for all $c \in C$ we get $D \subseteq L(C^{-1})$. Thus $DC \subseteq \subseteq L(C^{-1})C \subseteq P^{-1}$ and so $(P \supset)\ U(DC) \supseteq U(L(C^{-1})C) \supset P$, whence (i) follows.

---

[26] Note that the dot indicates the product in $G^{\#}$, while juxtaposition denotes the complex multiplication in G.

[27] Moreover, it is a (conditional) l. o. semigroup, since
$$(A \vee B) \cdot C = (A \cup B)^{\#} \cdot C = [(A \cup B) C]^{\#} = (AC \cup BC)^{\#} =$$
$$= (AC)^{\#} \vee (BC)^{\#} = A \cdot C \vee B \cdot C.$$

[28] Condition (i) has been given by EVERETT [1], (iii) in a special case by BAER [2]. — Note that $C^{-1} = [x^{-1} \mid x \in C]$.

The inclusion $u \in U(L(C^{-1})C)$ is equivalent to $u \geq xc$ for all $x \in L(C^{-1})$ and $c \in C$. This may be written in the form $c^{-1} x^{-1} \geq u^{-1}$ for $c^{-1} \in C^{-1}$, $x^{-1} \in L(C^{-1})^{-1} = U(C)$. Hence (ii) is equivalent to (i). Since $U(C)x \subseteq U(C)$ means that $c \leq ux$ for all $c \in C$, $u \in U(C)$, that is, $u^{-1} c \leq x$ for all $u^{-1} \in L(C^{-1})$, $c \in C$, the relation $U(C)x \subseteq U(C)$ is equivalent to $x \in U(L(C^{-1})C)$. This completes the proof.

The elements of $G^{\#}$ which have an inverse form a subgroup of $G^{\#}$. This will be called the *Dedekind extension* G_D of G.

When is G_D equal to $G^{\#}$? By the lemma of **9**, this can happen only if G is completely integrally closed. This condition is also sufficient, as is seen from the following theorem.

Theorem 19.[29] *A p. o. group G can be embedded in a complete p. o. group with preservation of order, g. l. b. and l. u. b., if, and only if, G is completely integrally closed. If G is directed, its Dedekind extension is a complete l. o. group.*

In order to prove sufficiency, let G be completely integrally closed. We verify that condition (iii) in the last lemma is fulfilled. $U(C)x \subseteq U(C)$ implies $U(C)x^n \subseteq U(C)$ for $n = 1, 2, \ldots$ whence $ux^n \geq c$, i. e. $c^{-1} u \geq x^{-n}$ for (all) $u \in U(C)$, $c \in C$. The hypothesis implies $x^{-1} \leq e$, i. e. $x \in P$. Thus $G^{\#}$ is a group. Since in $G^{\#}$ every set bounded from above has a l. u. b., the proof is finished.

Corollary 20. *A completely integrally closed directed group is commutative.*

This is an immediate consequence of Theorems 18 and 19.

11. The Cantor extension

The other most important method of obtaining the real numbers is CANTOR's method of fundamental sequences. It turns out, however, that by the Cantor process it is not always possible to complete a p. o. group even if it is l. o. and can be completed by the Dedekind process.

---

[29] There are many contributors to this theorem: KRULL [1], LOREN-ZEN [1], CLIFFORD [1], EVERETT—ULAM [1].

To analyze the Cantor sequence completion method, we use o-convergence introduced in **8** of Chapter II. In order to attribute a meaning to the product of Cantor sequences, it is necessary to restrict the sequences to those of a certain type connected with some ordinal. For the sake of simplicity we shall use— following EVERETT [1]—ordinary sequences a_n $(n = 1, 2, \ldots)$.[30]

Let G be a l. o. group. A sequence (u_n) is called a *unit sequence* if $u_n \to e$ in the sense of o-convergence.[31] A *fundamental sequence* (a_n) is one satisfying

$$\left| a_n a_m^{-1} \right| \leq u_n$$

for some unit sequence (u_n) and for all $m \geq n$. Here we may assume, without loss of generality, that $u_n \downarrow e$. Every o-convergent sequence is fundamental, and if the converse holds, G is said to be *o-complete*.

Under the obvious definition of product, $(a_n) \cdot (b_n) = (a_n b_n)$, the fundamental sequences form a group H in which the unit sequences form a normal subgroup E. Define the *Cantor extension*[32] G_C of G as the factor group H/E. From **4** I), O) we conclude that, together with (a_n) and (b_n), $(a_n \vee b_n)$ is again a fundamental sequence, and if $(a_n) \equiv (c_n)$ (mod E), then $(a_n \vee b_n) \equiv (c_n \vee b_n)$ (mod E).[33] Therefore the union $(a_n) \vee (b_n)$ of two fundamental sequences may be defined as $(a_n \vee b_n)$. Let $(a_n) \geq (b_n)$ mean that $(a_n \vee b_n) \equiv (a_n)$ (mod E); then the monotony laws hold. Consequently G_C is again a l. o. group. If (a) denotes the sequence with constant terms a, then we can state

Theorem 21. (EVERETT [1].) *The Cantor extension*[32] *G_C of a l. o. group G is a l. o. group. The canonical map $a \to (a)$*

---

[30] In general transfinite sequences must be used. We have to use types for which unit sequences with different terms exist, otherwise our considerations are trivial. For this problem in the f. o. case see COHEN—GOFFMANN [1].

[31] Equivalently, $|u_n| \leq v_n$ for some (v_n) with $v_n \downarrow e$.

[32] Strictly speaking, G_C is not completely determined by G. It depends on the underlying type of fundamental sequences. It is not left-right symmetric.

[33] $\left| (a_k \vee b_k)(a_m \vee b_m)^{-1} \right| \leq$
$\leq \left| (a_k \vee b_k)(a_k \vee b_m)^{-1} \right| \left| (a_k \vee b_m)(a_m \vee b_m)^{-1} \right| \left| (a_k \vee b_k)(a_k \vee b_m)^{-1} \right| \leq$
$\leq \left| b_k b_m^{-1} \right| \left| a_k a_m^{-1} \right| \left| b_k b_m^{-1} \right|.$

*of G into G_C is an o-isomorphism, and if corresponding elements
are identified, then every fundamental sequence in G o-converges
in G_C and every element of G_C is the o-limit of some fundamental
sequence in G.*

The final statement is evident, since the cosets mod E are
just the o-limits of the sequences lying in the cosets. Q. E. D.

The Cantor extension G_C of G is in general not o-complete.
To obtain a sufficient condition for the o-completeness of G_C,
let us call a l. o. group G *diagonal* if, given a double sequence
a_{kn} ($k, n = 1, 2, \ldots$) in G such that each sequence (a_{kn}) with
fixed k and varying n tends decreasingly to e, there exists
a sequence (b_k) with $b_k \to e$ in G such that $a_{kn_k} \leq b_k$ for suitable
choice of n_k.

Theorem 22. (EVERETT [1].) *The Cantor extension G_C of
a l. o. diagonal group G is o-complete (and again diagonal).*

Let us prove first that if $a_n, a \in G$ and $a_n \to a$ in G_C, then
$a_n \to a$ in G as well. Let $|\, a_n\, a^{-1}\,| \leq \gamma_n \downarrow e$ in G_C where[34]
$\gamma_n = (u_{nm})$, $|\, u_{nm}\, u_{nk}^{-1}\,| \leq v_{nm}$ ($k \geq m$) with $v_{nm} \downarrow e$ in G.
From diagonality we obtain the existence of a sequence (w_n)
such that $v_{nm_n} \leq w_n \downarrow e$ in G. Now

$$e \leq v_{nm_n}\, u_{nm_n}\, u_{nm}^{-1} \leq v_{nm_n}^2 \leq w_n^2 \quad \text{for } m \geq m_n;$$

and if we write $u_n = v_{nm_n}\, u_{nm_n}$, then $u_{nm} \leq u_n \leq w_n^2\, u_{nm}$
in G and $\gamma_n \leq u_n \leq w_n^2\, \gamma_n$ in G_C. Consequently,

$$e \leq \gamma_n = \gamma_1 \wedge \ldots \wedge \gamma_n \leq u_1 \wedge \ldots \wedge u_n \leq w_n^2\, \gamma_n,$$

and so $u_1 \wedge \ldots \wedge u_n \downarrow e$ in G_C. But if a monotone sequence
in G has the o-limit $e\, (\in G)$ in G_C, then it o-converges to e in
G, showing that $|\, a_n\, a^{-1}\,| \leq u_1 \wedge \ldots \wedge u_n \downarrow e$ and $a_n \to a$ in G.

Now assume that $a_k \in G_C$ ($k = 1, 2, \ldots$) is a fundamental
sequence in G_C, $|\, a_k\, a_l^{-1}\,| \leq \gamma_k \downarrow e$ ($l \geq k$). Plainly, $a_k = (a_{kn})$
satisfies $|\, a_{kn}\, a_{km}^{-1}\,| \leq u_{kn} \downarrow e$ in G ($m \geq n$). By diagonality
there exists a sequence $v_k \downarrow e$ in G such that $u_{kn_k} \leq v_k$ for
some n_k. Define $b_k = a_{kn_k}$; then $|\, a_k\, b_k^{-1}\,| \leq u_{kn_k} \leq v_k$. From
this and the inequalities $|\, a_l\, b_l^{-1}\,| \leq v_l \leq v_k$ ($l \geq k$) and

[34] In this paragraph convergence is to be understood with respect
to m.

$\left| a_k\, a_l^{-1} \right| \leq \gamma_k$ we obtain[35] $\left| b_k\, b_l^{-1} \right| \leq v_k\, \gamma_k\, v_k\, \gamma_k\, v_k \downarrow e$. By the preceding paragraph $b_k\, b_l^{-1}$ is o-convergent in G, i. e. $\left| b_k\, b_l^{-1} \right| \leq$ $\leq w_k \downarrow e$ in G $(l \geq k)$ and $(b_k) = a \in G_C$. Clearly, $\left| a\, b_k^{-1} \right| \leq$ $\leq w_k$ whence $\left| a_k\, b_k^{-1} \right| \leq v_k$ implies $\left| a a_k^{-1} \right| \leq w_k\, v_k\, w_k \downarrow e$. Thus $a_k \to a$ in G_C and G_C is o-complete.

Let (a_{kn}) be, for each fixed k, a sequence $\downarrow e$ in G_C. Then— as shown for γ_n in the last paragraph but one— $a_{kn} \leq u_{kn} \in G$ for some $(u_{kn}) \downarrow e$ in G, and the diagonality of G implies that of G_C. This completes the proof of Theorem 22.

Let G be a commutative l. o. group. We wish to get insight into the interrelation between the Cantor and Dedekind extensions of G. The next result furnishes us with most of the information desired.

Theorem 23. (EVERETT [1].) *Let G be a commutative l. o. group, G_D its Dedekind and G_C its Cantor extension. G_D is o-complete and contains G_C as a subgroup-sublattice.*[36]

Let C_n $(n = 1, 2, \ldots)$ be a fundamental sequence in G_D, i. e.[37]

$$(1) \qquad \left| C_n \cdot C_m^{(-1)} \right| \leq U_n \downarrow e \quad (m \geq n)$$

in G_D. Then $D_n = C_1 \wedge \ldots \wedge C_n$ is again a fundamental sequence, for $C_n \leq U_n \cdot C_{n+1}, \ldots, U_n \cdot C_{n+k}$ implies $C_n \leq$ $\leq U_n \cdot (C_{n+1} \wedge \ldots \wedge C_{n+k})$, and therefore

$$D_n \leq U_n \cdot D_n \wedge U_n \cdot (C_{n+1} \wedge \ldots \wedge C_{n+k}) = U_n \cdot D_{n+k}.$$

Now $D_n \cdot D_{n+k}^{(-1)} \leq U_n$ implies $U_1^{(-1)} \cdot D_1 \leq D_m$ for all m, i. e. D_m is bounded from below, and so $\wedge D_m = D$ exists in $G^{\sharp}$. This D satisfies $U_n^{(-1)} \cdot D \leq D \leq D_n$. Let $a \in L(U(D)D^{-1})$, $a \leq ud^{-1}$ for all $u \in U(D)$, $d \in D$. Because of $D \geq U_n^{(-1)} \cdot D_n =$ $= L(U_n^{-1}) \cdot D_n \supseteq L(U_n^{-1})D_n$ we have $a \leq u d_n^{-1} v_n^{-1}$ for all $d_n \in D_n$ and $v_n \in L(U_n^{-1})$, i. e. $a v_n \leq u d_n^{-1}$. Since $D_n \geq D$, $U(D_n) \subseteq U(D)$, we have $a v_n \leq u_n d_n^{-1}$ for all $u_n \in U(D_n)$, $d_n \in D_n$. Hence $D_n \in G_D$ implies $a v_n \leq e$. Therefore $a \leq v_n^{-1}$,

---

[35] Make use of **4** I).

[36] This result holds for every Cantor extension. Cf. footnote [32].

[37] $C^{(-1)}$ denotes the inverse of C in G_D and the dot indicates product in $G^{\sharp}$.

$a \in L(L(U_n^{-1})^{-1}) = L(U(U_n)) = U_n$, whence $U_n \downarrow e$ implies $a \leq e$. This establishes, by the dual of the lemma of **10**, that D belongs to G_D.

In order to prove that C_n has an o-limit in G_D, note that what has been proved shows that a fundamental sequence in G_D has an intersection in G_D. Hence $C_n \wedge \ldots \wedge C_{n+k} \wedge \ldots = E_n$ exists in G_D. Let $\vee E_n = E$; this lies in $G^\sharp$. Dually, $C_n \vee \ldots \vee C_{n+k} \vee \ldots = F_n$ exists in G_D. Let $\wedge F_n = F$ in $G^\sharp$. Evidently, $E \leq F$. From (1) we obtain $C_m \leq U_n \cdot C_n$ for all $m \geq n$ whence $C_{m+1} \leq U_{n+1} \cdot C_{n+1} \leq U_n \cdot C_{n+1}$, and more generally, $C_m \leq U_n \cdot C_k$ for $m \geq k \geq n$. Thus $E_m \leq U_n \cdot C_k$ for $k, m \geq n$ and $U_n^{(-1)} \cdot E_m \leq C_n \wedge \ldots \wedge C_{n+l} \wedge \ldots = E_n$, $E_m \cdot E_n^{(-1)} \leq U_n \downarrow e$ for $m \geq n$. This shows that (E_n) is a fundamental sequence, and the same holds for (F_n), consequently, E and F belong to G_D. Again from (1) we obtain $U_n^{(-1)} \cdot C_n \leq C_m \ (m \geq n)$, $U_n^{(-1)} \cdot C_n \leq E_n \leq E$, $C_m \leq U_m \cdot E \leq U_n \cdot E$ for $m \geq n$. Hence $F_m \leq U_n \cdot E$ and $F \leq U_n \cdot E$ so that $F \leq E$ and $E = F$. We conclude from $E_n \leq C_n \leq F_n$ that $C_n \rightarrow E$. This establishes the o-completeness of G_D.

Next let $a = (a_n)$ be a fundamental sequence in G. To show that it is fundamental in G_D as well, it suffices to verify that if $u_n \downarrow e$ in G, then $u_n \downarrow e$ in G_D. Let $C \in G_D$ satisfy $C \leq u_n^\sharp$ for all n, and let $c \in C$. Then $c \leq u_n$ in G and so $c \leq e, C \subseteq L(e)$. Now G_D being o-complete, (a_n) has an o-limit C_a in G_D. The correspondence $a \rightarrow C_a$ from G_C into G_D is easily seen to be one-to-one and order preserving. Q. E. D.

A very instructive analysis of the interrelation between topological completion and Dedekind completion has been given by BANASCHEWSKI [2].

12. Ideal systems

The transition from the elements of a ring to its ideals can also be regarded as a certain kind of completion process. This method has been studied extensively by H. PRÜFER[38] in domains of integrity and generalized by LORENZEN [1] to

---

[38] *Journ. reine u. angew. Math.*, **168** (1932), 1—36.

p. o. groups. The ideal systems are of great importance in p. o. groups; here we content ourselves with the discussion of their connection with completions of p. o. groups.

Let G be a directed group and

$$X \rightarrow X_r$$

a mapping from the set of all non-empty, l-bounded subsets X of G into the set of all subsets of G. We assume that it satisfies the following conditions:

I1. $X \subseteq X_r$,
I2. $X \subseteq Y_r$ implies $X_r \subseteq Y_r$,
I3.[39] $(a)_r = U(a)$,
I4. $aX_r = (aX)_r$ and $X_r a = (Xa)_r$ for all $a \in G$.

In this case we call the subsets X_r *r-ideals* and say that X is a generating system of X_r. The set $[X_r]$ of all r-ideals of G is called the (*total*) *r-ideal system* Σ_r of G.

Note that the system Σ_r completely determines the mapping $X \rightarrow X_r$, for I1 and I2 imply that a non-void intersection of r-ideals is likewise an r-ideal, and therefore X_r is equal to the intersection of all r-ideals containing X.

It is easily verified that I2 may be replaced by the following conditions:

I5. $X \subseteq Y$ implies $X_r \subseteq Y_r$,
I6.[40] $(X_r)_r = X_r$.

A *finite r-ideal* X_r is one generated by a finite subset X of G. The finite r-ideals form the finite r-ideal system of G.

Amongst the ideal systems of G there exist two extreme ones. Call the r-system *finer* than the q-system if, for all X, the relation $X_r \subseteq X_q$ holds.

Proposition 24. *In a directed group G, there exists a unique finest ideal system, the s-system defined by*

$$X_s = \bigcup_{x \in X} (x),$$

---

[39] Thus the (principal r-ideals) $(a)_r$ do not depend on the special choice of the ideal system. We may simply write $(a)_r = (a)$.

[40] X_r has the same lower bounds as X, for $X \subseteq (a)$ implies $X_r \subseteq (a)$.

and a unique coarsest ideal system, the v-system defined by[41]

$$X_v = \bigcap_{X \subseteq (x)} (x) = U(L(X)).$$

That is, for an arbitrary r-ideal system of G we have

$$X_s \subseteq X_r \subseteq X_v.$$

A straightforward calculation shows that the X_s and the X_v actually satisfy the requirements I1–4. If we consider an arbitrary r-ideal system, then $x \in X_r$ implies $(x)_s = (x)_r \subseteq X_r$, whence $X_s \subseteq X_r$. Also, if y is a lower bound of X, we have $X \subseteq (y)$, $X_r \subseteq (y)$, finally, $X_r \subseteq X_v$. Q. E. D.

Let us fix an arbitrary r-ideal system Σ_r of G. We define the union of two r-ideals X_r and Y_r as

(1) $$X_r \vee Y_r = (X \cup Y)_r$$

and their product as

(2) $$X_r \cdot Y_r = (XY)_r.$$

First of all we have to verify that these definitions are independent of the special choice of the generating systems X, Y. If $X_r \subseteq Z_r$ and $Y_r \subseteq Z_r$, then $X \subseteq Z_r$ and $Y \subseteq Z_r$ whence $X \cup Y \subseteq Z_r$ and $(X \cup Y)_r \subseteq Z_r$. This establishes the statement for the union. Turning to the product, we have

$$X_r \, Y = \bigcup_y X_r \, y = \bigcup_y (Xy)_r \subseteq (XY)_r,$$
$$(X_r \, Y)_r \subseteq (XY)_r, \quad (X_r \, Y)_r = (XY)_r.$$

Similarly we obtain $(X_r \, Y_r)_r = (X_r \, Y)_r$ whence

$$(XY)_r = (X_r \, Y_r)_r,$$

and the definition of products is legitimate. Clearly

$$(a)_r \cdot X_r = (aX)_r \quad \text{and} \quad X_r \cdot (a)_r = (Xa)_r.$$

It is readily seen that the union is commutative and associative, moreover, the rule (1) can be easily extended to any set

[41] This is actually the dual of the construction used to form Dedekind cuts.

of r-ideals $(X_\lambda)_r$ $(\lambda \in \Lambda)$ bounded from below: $\vee(X_\lambda)_r = (\cup X_\lambda)_r$. Multiplication (2) is obviously associative and satisfies

(3) $\quad X_r \leq Y_r$ implies $X_r \cdot Z_r \leq Y_r \cdot Z_r$ and $Z_r \cdot X_r \leq Z_r \cdot Y_r$.

The infinite distributive laws

(4) $\quad \vee(X_\lambda)_r \cdot Y_r = \vee((X_\lambda)_r \cdot Y), \quad Y_r \cdot \vee(X_\lambda)_r = \vee(Y_r \cdot (X_\lambda)_r)$

for a set of r-ideals (X_λ) bounded from below are immediate consequences of (1) and (2).

Proposition 25. *The r-ideals of a directed group G form a conditionally complete l. o. semigroup Σ_r containing the dual of G.*

The natural embedding of the dual of G is given by the map $a \rightarrow (a)$. This clearly reverses order and carries the l. u. b., if it exists, into the g. l. b., and conversely.

We want to find conditions ensuring that Σ_r will again be a group.

Lemma. *Assume that $X \cdot Y_r = (e)$ holds for two r-ideals X_r, Y_r. Then they are both v-ideals and*

$$Y_r = U(X^{-1}), \quad X_r = U(Y^{-1}).$$

$(XY)_r = (e)$ implies $XY \subseteq (e)$, and thus $Y \subseteq U(X^{-1})$. On the other hand, $u \in U(X^{-1})$ implies $uX \subset (e)$, $uX_r \subseteq (e)$, whence $uX_r \cdot Y_r \subseteq Y_r$ and $u \in Y_r$. Therefore $Y_r = U(X^{-1})$. But $U(X^{-1})_v = U(L(U(X^{-1}))) = U(X^{-1})$, as we wished to show.

We infer that the v-ideal system is the only ideal system which can form a group under multiplication (2). Repeating the arguments of Theorem 19, we are led to the result:

Theorem 26. *An ideal system Σ_r of a directed group G forms a group if, and only if, it is the v-ideal system and G is completely integrally closed. In this case Σ_v is a complete l. o. group containing the dual of G.*

We refer the reader to JAFFARD [10] for further results on the theory of ideal systems.

PARTIALLY ORDERED RINGS AND FIELDS

CHAPTER VI

PRELIMINARIES ON PARTIALLY ORDERED RINGS

1. Partial order on rings and fields

The term "ring" will be used for a ring which is not necessarily associative and not necessarily commutative; and "field" will mean an associative ring with division.

We shall say that R is a *partially ordered ring* ($p. o.$ *ring*) if

R1. R is a ring,

R2. R is a p. o. set under a relation $\leq$,

R3. $a \leq b$ implies $a + c \leq b + c$ for all $c \in R$,

R4. $a \leq b$ and $c > 0$ imply $ca \leq cb$ and $ac \leq bc$ ($a, b, c \in R$).[1]

Thus the additive group R_+ of R is a commutative p. o. group, hence it has the properties established for p. o. groups in general. If R_+ is f. o., directed or l. o., we call R a *f. o., directed* or *l. o. ring.*

$a \in R$ is called *positive* if $a \geq 0$, *negative* if $a \leq 0$. The set P of positive elements, the *positive cone* of R, uniquely determines the partial order of R, namely $a \leq b$ is equivalent to $b - a \in P$. The most important properties of P are incorporated in the following analogue of Theorem 2 of Chapter II.

Theorem 1. *A subset P of a ring R is the positive cone of some partial order of R if, and only if, the following conditions are fulfilled:*[2]

    a. $P \cap - P = 0$;

    β. $P + P \subseteq P$;

    γ. $PP \subseteq P$.

P defines a full order on R exactly if $P \cup - P = R$.

---

[1] A certain amount of complication, but at the same time considerable generality is introduced into the study of p. o. rings by postulating R4 instead of the more usual R4* (see below).

[2] Of course we mean by $-P$ the set $[-x \mid x \in P]$. Addition and multiplication of subsets of rings are to be understood as complex addition and complex multiplication.

Conditions β and γ assert that P is a semiring, while α tells us that P contains 0 but no other element along with its negative. Equivalently, a sum of elements of P is never 0 unless all terms are 0. For brevity, we shall call a semiring with this property a *conic semiring*.

It is an easy exercise to verify the rule of signs: the product of two negative elements is positive, the product of a positive and a negative element is negative, etc.

The additive group of a f. o. ring R must be torsion-free; hence the characteristic of R is 0.

If R possesses an identity e, then either $e > 0$ or $e \parallel 0$. In the second alternative it is possible to extend the partial order so as to have e positive unless R is of positive characteristic.[3]

The semirings which can be positive cones are characterized by the next

Theorem 2. *A semiring P with 0 is a positive cone of some p. o. ring if, and only if,*

    (a) *the additive semigroup of P is commutative and cancellative,*

    (b) *P is a conic semiring.*

Only the sufficiency needs a verification. By (a), the additive semigroup P_+ of P can be embedded in a minimal Abelian group R_+. The elements of R_+ are of the form $a - b$ with $a, b \in P_+$. It is routine to check that if multiplication is defined by the distributive rule $(a - b)(c - d) = (ac + bd) - (ad + bc)$, then R_+ becomes a ring R containing P as a subsemiring. In view of (b), P defines in fact a partial order on R.

We shall have frequent occasion to consider rings without divisors of zero. In such rings R4 can be replaced by the stronger condition

   · R4*. $a < b$ and $c > 0$ imply $ca < cb$ and $ac < bc$.

In general, partial orders satisfying R4* will be called *strict*. We have immediately:

Lemma. *A necessary and sufficient condition for a partial*

---

[3] The semiring generated by the positive cone and e is conic except when $nx = 0$ holds for all $x \in R$ and some natural integer n; in this case P must be 0.

order P of a ring to be strict is that P be a semiring without divisors of zero.

In fact, R4* is R4 together with the property that $c(b - a)$ [and $(b - a)c$] is never 0 unless $b - a$ ($\in P$) or c ($\in P$) vanishes.

A f. o. ring R obviously satisfies

R5. if $ab > 0$ and one of a, b is > 0, then so is the other.

If a p. o. ring R satisfies R5, we shall say that its partial order is *division-closed*. If R is a division ring (i. e. a ring with division), it is sometimes useful to assume R5. This is equivalent to the fact that the positive cone P is a division semiring, and if associativity is also assumed, i. e. R is a field, then $P\backslash 0$ is a p. o. group.

There is no need for entering into the discussion of induced partial orders on subrings, convexity of subrings, o-homomorphisms etc., since it is essentially the same as that given for the corresponding notions in p. o. groups.

2. Examples

1) The ring of rational integers under the customary ordering is a f. o. ring. All the rational numbers and all the real numbers form f. o. fields under the usual ordering.

2) An additive commutative p. o. group (or f. o. group) under the trivial multiplication is a p. o. ring (f. o. ring).[4]

3) The polynomial ring $R[x]$ over an associative p. o. ring R without divisors of zero is made into a p. o. ring if one puts $f > 0$ ($f \in R[x]$) whenever the leading coefficient of f is > 0 in R. This is the lexicographic ordering of $R[x]$. Another possibility is to put $f > 0$ if the coefficient of the lowest nonvanishing term is > 0 — this may be called the antilexicographic ordering. In both cases $R[x]$ is f. o. if, and only if, R is f. o.

The polynomial ring $F[x]$ over a subfield F of the real numbers can also be f. o. by defining $f > 0$ if $f(\theta) > 0$ where θ is a fixed real number transcendental over F.

---

[4] That is, all products are zero. Such rings are called *zero-rings*.

4) The field of rational functions with coefficients in a commutative f. o. field F can be f. o. by defining $fg^{-1} \geq 0$ with $f, g \in F[x]$ if $fg \geq 0$ in the lexicographic ordering of $F[x]$.

5) The real-valued continuous functions on the real interval $[0, 1]$ form a l. o. ring if we set $f \geq 0$ whenever $f(x) \geq 0$ for all x.

More generally, the same holds if we replace the interval $[0, 1]$ by a topological space.

6) The $n \times n$ square matrices with elements in a p. o. ring R form a p. o. ring by defining a matrix $\| a_{ik} \|$ to be positive if each a_{ik} is positive. If R is l. o., the same will be true for the matrix ring.

7) The last example may be generalized by taking an algebra A (of finite or infinite rank) over an associative p. o. ring R such that A has a basis e_i $(i \in I)$ with positive structure constants:

$$e_i e_j = \sum_k \gamma_{ij}^k e_k \qquad (\gamma_{ij}^k \in R, \ \gamma_{ij}^k \geq 0).$$

Positivity is defined by

$$\sum a_i e_i \geq 0 \ (a_i \in R) \text{ if } a_i \geq 0 \text{ for every } i.$$

A is a l. o. ring if R is a l. o. ring.

As a special case we mention the group algebra of a group over an associative f. o. ring.

8) The group algebra A of a f. o. group G over an associative f. o. ring R admits a linear order preserving those of G and R. One defines

$$\sum_{i=1}^{n} a_i g_i > 0 \qquad (0 \neq a_i \in R, \ g_i \in G, \ g_1 < \ldots < g_n)$$

if $a_n > 0$.

9)[5] For every f. o. field F and for an arbitrary natural number n, a nilpotent f. o. algebra A of rank n over F can be defined by taking a symbol c and then defining A as an algebra

---

[5] See ZEMMER [1].

over F with a basis $c, c^2, \ldots, c^n$; the basis elements are multiplied as powers of c subject to the condition $c^{n+1} = 0$. We set

$$\sum \lambda_i c^i > 0 \quad (\lambda_i \in F)$$

if the first non-zero λ_i is > 0.

10)[6] Consider the ring $\mathfrak{E}(G)$ of all endomorphisms of a directed Abelian group G, and define an endomorphism θ positive if it carries the positive cone P of G into itself. Then $\mathfrak{E}(G)$ will be a p. o. ring. (Note that only in case of directed groups can we conclude that if both θ and $-\theta$ are positive, then $\theta = 0$; in fact, $\theta, -\theta \geq 0$ imply $g\theta = 0$ for all $g \in P$, i. e. θ is zero on $\{P\}$.)

Instances of f. o. non-commutative fields are the formal power series fields, see Chapter VIII, **5.**

3. Ordering of rings of quotients

In this section we shall be concerned with associative rings only.

Every partial order P of a ring R defines a partial order on a ring S containing R, namely, that with the same positive cone. It is much less trivial to continue a given full order P of R to a full order of a ring S of quotients of R. The classical method of extending the ordering relation of the integers to the rational numbers is applicable in a great many cases.

Let S be a ring containing R and assume that to each $a \in S \setminus R$ there exist elements $a, b \in R$ such that

   (i) a is not a left divisor of zero in S;

   (ii) b is not a right divisor of zero in S;

   (iii) $aa = c$ and $ab = d$ belong to R.

In this case S will be said to be *a ring of quotients of R.*

Theorem 3. (FUCHS [11].) *A full order P of an associative ring R can be uniquely extended to a full order Q of an arbitrary ring S of quotients of R.*

[6] See BIRKHOFF—PIERCE [1].

If R is f. o., then in the above definition we may take $a, b > 0$. Then c and d have the same sign, for $cb = aab = = ad$, and so the sign of aa and ab does not depend on the special choice of a or b. Define a to belong to Q if either $a \in P$ or $a \in S \setminus R$ and aa (and so ab) lies in P. Then we know already that $Q \cap -Q = 0$. If $a, \beta \in Q$, then there exist elements $a, b \in P$ or $= 1$ such that $aa, \beta b \in P$. Hence $a(a + \beta)b = = (aa)b + a(\beta b) \in P$, and so $a + \beta \in Q$.[7] Again, $a(a\beta)b = = (aa)(\beta b) \in P$ whence $a\beta \in Q$. Consequently, Q is a positive cone, and since obviously $Q \cup -Q = S$, Q defines a full order on S. The uniqueness is evident.

The following immediate consequences deserve mention.

Corollary 4. (ALBERT [2], NEUMANN [2].) *If R is an associative f. o. ring which has a field S of quotients (in the sense of* ORE),[8] *then S can be fully ordered uniquely so as to continue the ordering of R.*

Corollary 5. *A full order of a domain of integrity*[9] *can be uniquely extended to a full order of its field of quotients.*

A further useful result which can be derived from the above discussion is as follows:

Corollary 6. (GRÄTZER—SCHMIDT [1].)[10] *Let R be an associative ring and I an ideal of R containing an element which is no left divisor of zero and an element which is no right divisor of zero in R. Then every full order of I admits a unique extension to a full order of R.*

Note that R is a ring of quotients of I (with universal a and b in the definition).

---

[7] Here we make use of the fact that if $a, b > 0$ and $c = aab \in R$, then a and c have the same sign. This is indeed true, for if $d > 0$ is chosen so that $d(aa) = f \in R$, then the elements a, f, $fb = dc$, c are simultaneously positive or negative. (a, d are not left, b is not right divisor of zero.)

[8] *Annals Math.*, **32** (1931), 463—477.

[9] By a *domain of integrity* we mean a commutative and associative ring without divisors of zero. Corollary 5 is a familiar result contained in most textbooks on algebra.

[10] Corollaries 6 and 7 have been stated only for associative rings with no divisors of zero.

4. Embedding in a ring with identity

It is known that every ring without identity can be embedded in a ring with identity. The corresponding question for f. o. rings has, in general, a negative answer. The following corollary provides us with a criterion for the existence of full order in ring extensions with identity.

Corollary 7. (RÉDEI [1].) *Let R be an associative ring containing at least one not left and one not right divisor of zero and R^* a minimal ring with identity containing R. Every full order of R can be extended uniquely to a full order of R^*.*[11]

It is known that if the ring contains elements which are not divisors of zero, then it has a ring extension with identity in which they remain non-divisors of zero.[12] The term "minimal" is used to denote this ring extension. Our result follows at once from Corollary 6.

Note that the hypothesis of the presence of non-divisors of zero cannot be dropped in Corollary 7. In fact, let R be the set of all ordered pairs (m, n) of rational integers such that 1. equality and addition are defined component-wise, 2. multiplication is given by $(1, 0) (m, n) = = (m, n)$ and $(0, 1) (m, n) = (0, 0)$, 3. ordering relation is defined lexicographically. Then R is a f. o. ring. It cannot be embedded o-isomorphically in any f. o. ring R^* with identity e, since
$[(2, 0) -e] \cdot (1, 0) = (1, 0) > 0$ and $(0, 1) \cdot [(2, 0) -e] = -(0, 1) < 0$
imply that in R^* we must have both $(2, 0) > e$ and $(2, 0) < e$. (D. G. JOHNSON [1].)

It is natural to raise the following question: when is it possible to embed a f. o. ring as a convex ideal in a ring with identity? A necessary and sufficient condition may easily be given:

Theorem 8. (D. G. JOHNSON [1].) *A f. o. ring R without identity can be embedded as a convex ideal in a ring R^* with identity if, and only if,*

(1) $$ab \leq \min (a, b)$$

holds for all positive $a, b \in R$.

----

[11] Thus R is an O-ring (in the sense of the next section) if, and only if, R^* is one.

[12] See e. g. B. BROWN and N. H. McCOY, *Duke Math. Journ.*, **13** (1946), 9—20, or J. SZENDREI, *Acta Sci. Math. Szeged*, **13** (1950), 231—234.

The condition is necessary, for if $e \in R^*$ is the identity, then clearly $a < e$ for every positive $a \in R$ whence $ab \leq eb = b$ and similarly $ab \leq a$ for all $b \geq 0$. Conversely, if (1) is fulfilled by R, then consider the set R^* of all pairs (m, a) where m is an integer and $a \in R$ under the following rules: 1. equality and addition are defined component-wise, 2. multiplication is given by $(m, a) (n, b) = (mn, na + mb + ab)$, 3. we put $(m, a) \geq 0$ if $m > 0$ or if $m = 0$ and $a \geq 0$. It is straightforward to check that R^* is a f. o. ring with identity $(1, 0)$ containing R as a convex ideal. (Condition (1) ensures that if (m, a), $(0, b)$ are positive, then so are $(m, a) (0, b) = (0, mb + ab)$ and $(0, b) (m, a) =$ $= (0, mb + ba)$.)

CHAPTER VII

EXTENSIONS OF PARTIAL ORDERS IN RINGS

1. Extension to a full order; O-rings

Let $a_1, \ldots, a_n$ denote elements of an arbitrary ring R, A a subset of R, and $H(A, a_1, \ldots, a_n)$ the semiring generated by $0, A, a_1, \ldots, a_n$ in R. Our subsequent discussion rests on the following result.[1]

Theorem 1. (FUCHS [9].) *A partial order P of a ring R can be extended to a linear order of R if, and only if, P satisfies:*

(*) *for every finite set $a_1, \ldots, a_n \in R$ one can choose $\varepsilon_1, \ldots, \varepsilon_n$ $(\varepsilon_i = 1 \text{ or } -1)$ such that*

$$H(P, \varepsilon_1 a_1, \ldots, \varepsilon_n a_n)$$

is conic.

If P has a linear extension Q, then choosing ε_i such that $\varepsilon_i a_i \geq 0$ in Q we see that $H(P, \varepsilon_1 a_1, \ldots, \varepsilon_n a_n)$ is a subsemiring of Q, and hence it is conic.

The proof of sufficiency is based on the following

Lemma. *If a partial order P of R has property (*), then for each $x \in R$ either $H(P, x)$ or $H(P, -x)$ defines a partial order P' of R again with property (*).*

If both $H(P, x)$ and $H(P, -x)$ fail to satisfy (*), then there exist elements $a_1, \ldots, a_n$ and $b_1, \ldots, b_m$ such that for arbitrary choice of the signs $\varepsilon_1, \ldots, \varepsilon_n, \eta_1, \ldots, \eta_m$

$$H(P, x, \varepsilon_1 a_1, \ldots, \varepsilon_n a_n) \text{ and } H(P, -x, \eta_1 b_1, \ldots, \eta_m b_m)$$

are not conic. But then

$$H(P, \varepsilon x, \varepsilon_1 a_1, \ldots, \varepsilon_n a_n, \eta_1 b_1, \ldots, \eta_m b_m)$$

is conic for no choice of the signs $\varepsilon, \varepsilon_1, \ldots, \varepsilon_n, \eta_1, \ldots, \eta_m$, contrary to the hypothesis that P satisfies (*). Now put $P' = H(P, x)$ or $P' = H(P, -x)$ according as $H(P, x)$ or $H(P, -x)$ satisfies (*). Then P' will be a conic semiring with (*).

---

[1] Cf. the analogous Theorem 1 in Chapter III.

The proof of Theorem 1 can be now easily completed. By ZORN's lemma, the set of all extensions of P satisfying (*) contains a maximal member Q. By the lemma, for every $x \in R$, either $H(Q, x)$ or $H(Q, -x)$ belongs to the set considered, and so, by maximality, either $x \in Q$ or $-x \in Q$, that is, Q is a full order. Q. E. D.

Call a ring R an *O-ring* if it can be f. o. If we start with the trivial order $P = 0$, we obtain

Corollary 2. (PODDERYUGIN [1].) *A ring R is an O-ring if, and only if, for every finite set $a_1, \ldots, a_n$ in R it is possible to select signs $\varepsilon_i = 1$ or -1 such that $H(\varepsilon_1 a_1, \ldots, \varepsilon_n a_n)$ is conic.*

As a consequence of this corollary we get

Corollary 3. *If every finitely generated subring of R is an O-ring, then R is an O-ring too.*

The striking analogy between p. o. groups and rings breaks down, however, in several cases. For instance, the direct sum of O-rings is, in general, not an O-ring. Moreover, we have

Proposition 4.[2] *A direct sum of O-rings is again an O-ring if, and only if, all the components with at most one exception are zero-rings.*

For the necessity we show: if R is a f. o. ring and $R = R_1 \oplus R_2$ (direct sum in the pure ring-theoretical sense), then either R_1 or R_2 is a zero-ring. Suppose that R_1 is not a zero-ring. Then there exist elements $a, b \in R_1$ such that $ab \neq 0$ where we may assume $a, b > 0$. If $x > b$ for some $x \in R_2$, then $a(x - b) = -ab$ is a contradiction, for the left member is ≥ 0 and the right member is < 0. Hence $x < b$ for all $x \in R_2$. If $x, y \in R_2$ and, say $x, y > 0$, then $0 \leq (b - x)y = -xy \leq 0$ implies $xy = 0$, i. e. R_2 is a zero-ring.

Conversely, if R_λ ($\lambda \in \Lambda$) is a set of O-rings all of which except for one R_λ, say R_1, are zero-rings, then the direct sum $\sum R_\lambda$ is likewise an O-ring. Well-order the index set Λ beginning with 1 and take full orders on the R_λ. It is easy to check that if one defines the ordering lexicographically, one arrives at a full order on $\sum R_\lambda$.

---

[2] The essential part of this result is due to ZEMMER [1]. This result is true for complete direct sums as well.

Here it is the proper place to state:

Proposition 5. (BIRKHOFF—PIERCE [1].) *An associative semi-simple ring[3] is an O-ring if, and only if, it is an O-field.*

This follows from the more general result: a full matrix ring of degree $n \geq 2$ over any ring R containing an element a with non-zero square is not an O-ring. In fact, if e_{ik} is the matrix with 1 in the (i, k)-place and 0's elsewhere, then

$$(ae_{11} + ae_{22})^2 + (ae_{21} - ae_{12})^2 = 0$$

which is impossible in an O-ring. Hence $n=1$ and the celebrated theorem of WEDDERBURN — ARTIN, together with the preceding proposition, completes the proof. (Another proof may be given by making use of Theorem 6 of Chapter VIII.)

If we start with an algebra, then we obtain similarly:

Proposition 6. *An associative semi-simple O-algebra over a commutative field is an O-field.*

Moreover, the O-field must be commutative (ALBERT [1]). For if a is an element of the O-field F not in its centre C, then it satisfies an equation $f(x) = x^m + a_1 x^{m-1} + \ldots + a_m = 0$ with $m > 1, a_i \in C$. Now $\beta = a - m^{-1}a_1 \in F$ is a root of an equation $g(y) = y^m + b_2 y^{m-2} + \ldots + b_m = 0$ with $b_i \in C$. If this is of minimal degree, then by a theorem of WEDDERBURN[3a]

$$g(y) = (y - a_1) \ldots (y - a_m)$$

for conjugates a_i of a. But in an O-field the sum of conjugates of a non-zero element cannot vanish.

2. *O*-rings without divisors of zero

The rings without divisors of zero are of particular interest. We begin their discussion with the rather elementary

Lemma. *A ring R admits a strict full order if, and only if, it is an O-ring containing no divisors of zero.*

---

[3] By a semi-simple ring we mean one containing no non-zero nil-potent ideals and satisfying the minimum condition for left ideals. Such rings are direct sums of finitely many complete matrix rings over fields.

[3a] Cf. L. E. DICKSON, *Algebras and their arithmetics* (Chicago, 1923).

We have seen that a partial order is strict if, and only if, its positive cone is a semiring without divisors of zero. In the f. o. case this is equivalent to the absence of divisors of zero in R.

The main result is

Theorem 7. (FUCHS [9].) *Let P be a partial order of a ring R containing no divisors of zero. A necessary and sufficient condition for P to be extendible to a full order of R is that, for every finite set of elements $a_1, \ldots, a_n$ of R ($a_i \neq 0$), no sum of products containing each a_i an even number of times and (possibly an arbitrary number of times) elements $\neq 0$ of P as factors may vanish.*

Let $x_1, \ldots, x_s$ ($s \geq 2$) be non-zero products of the stated kind, built up from elements $a_1, \ldots, a_n$ of R ($a_i \neq 0$), and $x_1 + \ldots + x_s = 0$. Then because of $x_1, \ldots, x_s \in H(P, \varepsilon_1 a_1, \ldots, \varepsilon_n a_n)$ for all choices of $\varepsilon_1, \ldots, \varepsilon_n$, P does not satisfy (*) of Theorem 1. Thus the condition is necessary.

Conversely, assume that the stated condition holds. We show that the negation of (*) in Theorem 1 leads to a contradiction. We pick out from $H(P, \varepsilon_1 a_1, \ldots, \varepsilon_n a_n)$ elements h_j ($j = 1, \ldots, 2^n$) such that, for every choice of the signs $\varepsilon_1, \ldots, \varepsilon_n$, one of them is a 0 sum with non-zero terms.[4] We form the product $x = h_1 h_2 \ldots h_{2^n}$ after putting, say, $\varepsilon_1 = \ldots = \varepsilon_n = 1$. x is again a 0 sum with non-zero terms and it remains zero if arbitrary a_i are replaced by $-a_i$. Write $x = x_0 + x_1$ where x_0 (x_1) is the sum of terms containing a_1 an even (odd) number of times; evidently, either x_0 or x_1 or both exist. Replacing a_1 by $-a_1$, we have $x_0 - x_1 = 0$ which together with $x_0 + x_1 = 0$ yields $2x_0 = 0$. Now we start with $2x_0$ rather than x if x_0 exists or with x_1^2 if x_0 fails to exist and use the same procedure to eliminate the terms in which a_2 occurs an odd number of times, etc. Finally, we arrive at a 0 sum with non-zero terms, each term containing each one of $a_1, \ldots, a_n$ an even number of times. The resulting contradiction shows that (*) holds, and so P is extendible to a linear order. Q. E. D.

---

[4] Each h_j is regarded as a function of $\varepsilon_1, \ldots, \varepsilon_n$.

Calling a product containing each factor a_i ($\neq 0$) an even number of times *even*, we have

Corollary 8. (R. E. JOHNSON [1], PODDERYUGIN [1].) *A ring without divisors of zero is an O-ring if, and only if, no sum of even products vanishes.*[5]

Turning to fields, we have

Corollary 9. (FUCHS [9].) *Let P be a partial order of a field F. A necessary and sufficient condition that P have a linear extension in F is that no sum of terms of the form*

$$a_1^2 \ldots a_n^2 \quad or \quad pa_1^2 \ldots a_n^2 \quad (0 \neq a_i \in F,\ 0 \neq p \in P)$$

vanishes.

In fact, by the associative law and the existence of inverses we can write $aba = b(b^{-1})^2 (ba)^2$ whence each term indicated in Theorem 7 can be written in the form $a_1^2 \ldots a_n^2$ or $pa_1^2 \ldots a_n^2$.

If F is commutative, then we have

Corollary 10. (SERRE [1].) *A partial order P of a commutative field F is extendible to a full order of F if, and only if, the sum of elements of the form*

$$a^2 \quad or \quad pa^2 \quad (0 \neq a \in F,\ 0 \neq p \in P)$$

is never 0.

Call a field F *formally real* if -1 cannot be represented as a sum of products $a_1^2 \ldots a_n^2$ ($a_i \in F$). Thus a commutative field is formally real if, and only if, -1 is not a sum of squares.[6] We get the most important result:

Corollary 11. (ARTIN—SCHREIER [1], PICKERT [1], SZELE [1].) *A field can be fully ordered if, and only if, it is formally real.*[7]

---

[5] Corollary 8 does not remain valid if the hypothesis of absence of divisors of zero is omitted. (E. g. a ring having a zero-ring as a direct summand.)

[6] The fundamental notion of formally real commutative fields has been introduced by ARTIN—SCHREIER [1].

[7] This result has been proved in the commutative case by ARTIN—SCHREIER and then generalized to the non-commutative case independently by PICKERT and SZELE. For commutative O-fields cf. also DIEUDONNÉ [2]. Relations between f. o. fields and fields with valuations have been studied by BAER [1], KRULL [1] and CONRAD [2].

Take $P = 0$ in Corollary 9 and observe that $a_1^2 \ldots a_n^2 +$ $+ b_1^2 \ldots b_m^2 + \ldots = 0$ $(a_i \neq 0)$ may be written as

$$b_1^2 \ldots b_m^2 (a_n^{-1})^2 \ldots (a_1^{-1})^2 + \ldots = -1.$$

Let us understand by an O^*-ring a ring in which every partial order can be extended to a full order. It is easy to prove:

Proposition 12. *A ring R is an O^*-ring if, and only if, for each pair of finite sets of elements of R, $a_1, \ldots, a_n$ and $b_1, \ldots, b_m$, the fact that $H(a_1, \ldots, a_n)$ is conic implies that $H(a_1, \ldots, a_n, \varepsilon_1 b_1, \ldots, \varepsilon_m b_m)$ is conic for an appropriate choice of the signs ε_i.*

The necessity follows from the extendibility of $P = H(a_1, \ldots, a_n)$, while the sufficiency may be proved by observing that the semiring $H(P, b_1, \ldots, b_m)$ is conic if, and only if, $H(a_1, \ldots, a_n, b_1, \ldots, b_m)$ is conic for all finite subsets $a_1, \ldots, a_n$ of a partial order P.

It is easily inferred that if we go so far as to assume the ring to be a field, the condition of the last Proposition reduces to: If $H(a_1, \ldots, a_n)$ is conic, then so is $H(a_1, \ldots, a_n, b_1^2, \ldots, b_m^2)$ for every finite set of elements $b_1, \ldots, b_m$ in the field.

O-rings with divisors of zero will be studied in the next Chapter.

3. Real closed commutative fields

Particular attention must be devoted to the case of c o m- m u t a t i v e fields. In this case the theory culminates in the Artin—Schreier theory of real closed fields which we are going to discuss.[8]

First, we consider the problem of extending the full order of a commutative field K to a full order of an algebraic extension of K.

Proposition 13. (SERRE [1].) *If K is a f. o. commutative field and f is an irreducible polynomial over K such that*

$$f(a) f(b) < 0 \quad \text{for some } a, b \in K,$$

then the field L obtained from K by adjoining a root of f can be f. o. preserving the ordering of K.

We use an induction on the degree n of f. The statement being trivial for $n = 1$, assume it true for polynomials of degree $< n$ $(n \geq 2)$. If L had no full order continuing that

---

[8] For an elegant discussion of real closed fields we refer to BOUR- BAKI [1]; cf. also VAN DER WAERDEN [1].

of K, then, by Corollary 10, we could deduce the existence of a relation[9]

$$(*) \qquad \sum_{i=1}^{m} p_i f_i^2 = -1 + gf \quad (0 < p_i \in K)$$

with polynomials f_i, g in $K[x]$. Without loss of generality we may suppose that the degree of each f_i is less than, or equal to, $n - 1$, whence g is of degree $\leq n - 2$. K is formally real, thus by (*) we have $g(a) f(a) > 0$ and $g(b) f(b) > 0$, so that $g(a) g(b) < 0$. Therefore if we decompose g into irreducible factors over K, one of these factors, say h, must satisfy $h(a) h(b) < 0$. Thus we can use the induction hypothesis to conclude that the field obtained by adjoining to K a root of $h(x)$ is formally real. This shows that $\sum p_i f_i^2 \equiv -1 \pmod{h}$ is impossible. Hence (*) leads to a contradiction and the assertion follows.

Corollary 14. *If K is a f.o. commutative field and $a \in K$, $a > 0$, then $K(\sqrt{a})$ can again be f.o. preserving the ordering of K.*

If $f(x) = x^2 - a$ is irreducible, then in view of $f(0) f(a + 1) < 0$, the preceding proposition is applicable.

Theorem 15. (ARTIN—SCHREIER [1].) *If K is a commutative O-field and L is an algebraic extension of odd degree over K, then every full order of K can be extended to a full order of L.*

An O-field is of characteristic 0, hence L is separable over K, and we have $L \cong K[x]/(f)$ for some irreducible $f = x^n (1 + a_1 x^{-1} + \ldots + a_n x^{-n}) \in K[x]$ of odd degree n. If $m = \max(1, |a_1| + \ldots + |a_n|)$, then for $|x| > m$

$$\left| a_1 x^{-1} + \ldots + a_n x^{-n} \right| < \left| a_1 \right| m^{-1} + \ldots + \left| a_n \right| m^{-1} \leq 1$$

and $1 + a_1 x^{-1} + \ldots + a_n x^{-n} > 0$. We obtain $f(-2m) f(2m) < 0$ and Proposition 13 concludes the proof.

A maximal formally real commutative field, i. e. a formally real commutative field, no proper algebraic extension of which is formally real, is called *real closed*. The most important characterization is contained in

---

[9] The full order of K is division-closed, therefore from a zero sum of squares we can obtain a representation (*) of -1 in the same manner as in the proof of Corollary 11.

Theorem 16. (ARTIN—SCHREIER [1].) *For a f. o. commutative field F the following statements are equivalent*:

(a) $F(i)$ *is algebraically closed* $(i = \sqrt{-1})$;

(b) F *is real closed*;

(c) *every positive element of F has a square root in F and every polynomial of odd degree over F has a root in F.*

If (a) holds, then F has only one algebraic extension, namely $F(i)$. This is not real closed, since -1 is a square in it. Hence (b) follows.

If (b) holds, then, by Corollary 14, every positive element has a square root in F, and by Theorem 15 every polynomial of odd degree over F must have a root in F. Thus (c) holds.

If (c) is assumed for F, then take $F(i)$. First, we show that every quadratic polynomial over $F(i)$ has roots in $F(i)$. It is obvious that it suffices to verify that every element $a + bi$ $(a, b \in F)$ is a square in $F(i)$. But this can be done in the same manner as it is customary for complex numbers, because positive elements of F have square roots in F:

$$\sqrt{a + bi} = \sqrt{\tfrac{1}{2}\left(a + \sqrt{a^2 + b^2}\right)} + i\sqrt{\tfrac{1}{2}\left(-a + \sqrt{a^2 + b^2}\right)} \in F(i).$$

What remains to be proved is nothing else than the following

Lemma. *If F is a commutative field such that*

(i) *every polynomial of odd degree over F has a root in F,*

(ii) *every quadratic polynomial over $F(i)$ has a root in $F(i)$, then $F(i)$ is algebraically closed.*

We show that every polynomial f over F has a root in $F(i)$. This will suffice, for if g is a polynomial over $F(i)$ and $\bar{g}$ is obtained from g by replacing i by $-i$ in every coefficient, then $f = g\bar{g}$, being a polynomial over F, has a root $a + bi$ $(a, b \in F)$ in $F(i)$. Clearly, either $a + bi$ or $a - bi$ is a root of g.

Now if $f \in F[x]$ is of degree $n = 2^l q$ with odd q, then in case $l = 0$, (i) guarantees that f has a root in $F(i)$. Thus we may proceed by induction on l. It is proved in commutative field theory that there exists a finite algebraic extension L of F in which f splits into linear factors, that is to say,

$f = a(x - a_1) \ldots (x - a_n)$ with $a \in F$, $a_j \in L$. The $\binom{n}{2}$ elements

$$\beta_{jk} = a_j + a_k - c a_j a_k \quad (1 \leq j < k \leq n)$$

with fixed $c \in F$ are roots of a certain polynomial h over F. Since the degree of h is of the form $\binom{n}{2} = 2^{l-1} q'$ with odd q', we infer that one of the β_{jk} must belong to $F(i)$. This holds for every $c \in F$. As F is infinite (otherwise it could not satisfy (i)), we can find an index pair (j, k) such that both

$$a_j + a_k - c_1 \, a_j \, a_k \in F(i) \quad \text{and} \quad a_j + a_k - c_2 \, a_j \, a_k \in F(i)$$

for different elements c_1, c_2 of F. But then $a_j + a_k$ and $a_j \, a_k$ are elements of $F(i)$, and so a_j, a_k are roots of a quadratic polynomial $t \in F(i)[x]$. Assumption (ii) and the fact that f, t have common roots complete the proof of the lemma and hence that of Theorem 16.

Corollary 17. *A real closed commutative field admits one, and only one, full order.*

If a is an arbitrary element of the real closed commutative field F, then by virtue of Theorem 16 it has a square root $b + ci$ ($b, c \in F$) in $F(i)$. But $a = b^2 - c^2 + 2bci$ implies $bc = 0$, i. e. either $b = 0$ or $c = 0$, proving that one of $-a$, a is a square in F. The product of squares is again a square, and since a sum of squares cannot be the negative of a square, it is again a square. We see that the set P of all squares is a conic sub-semiring in F with $P \cup - P = F$. Thus P defines a full order on F which is plainly the only one on F.

A most essential result is:

Theorem 18. (ARTIN—SCHREIER [1].) *Every formally real commutative field K can be embedded in a real closed commutative field F which is algebraic over K. This F is, up to isomorphisms over K, uniquely determined by K.*

Let L be the algebraic closure of K, i. e. the least algebraically closed commutative field containing K. Consider all formally real subfields of L containing K. This set is non-void and is evidently inductive, hence we may apply ZORN's lemma to deduce the existence of a maximal member F in this set.

Thus F has no formally real proper algebraic extension, i. e. it is real closed. Clearly, F is minimal in the sense that no proper subfield containing K is real closed.[10]

In order to prove the isomorphy over K of two real closed algebraic extensions F and F^* of K, we consider the set of all pairs $[U, \theta]$ where U is a formally real subfield of F containing K and θ is an isomorphism over K mapping U onto a subfield U^* of F^*. We set $[U_1, \theta_1] \leq [U_2, \theta_2]$ if U_1 is a subfield of U_2 and θ_2 agrees with θ_1 on U_1. The resulting p. o. set is not void (for $[K, \iota]$ with the identity map ι belongs to it) and is inductive, therefore there exists a maximal element $[V, \chi]$ in this set. Assume that $V \neq F$. Then V must have a proper algebraic extension W, say by a root a of the irreducible polynomial $f \in V[x]$, which is formally real. χ being an isomorphism, the extension of $V^* = \chi(V)$ by a root $a^* (\in F^*)$ of the corresponding polynomial $f^* = \chi(f)$ is again formally real. As is shown in field theory, χ can be extended to an isomorphism χ' of $W = V(a)$ onto $W^* = V^*(a^*)$ such that a goes into a^*. This argument shows that the pair $[W, \chi']$ is a greater element in the set considered. The resulting contradiction implies that $V = F$, whence $V^* = F^*$ and the theorem is completely proved.

A similar argument may be applied to prove that if A is an algebraically closed commutative field containing a formally real field K, then there exists a real closed subfield F of A which contains K and satisfies $A = F(i)$. Hence an algebraically closed commutative field A of characteristic 0 contains a real closed subfield F such that $A = F(i)$.

The complex number field contains real closed fields distinct from, but isomorphic to, the real field; e. g. there is such a field containing $i\sqrt[4]{2}$ or $i\pi$.

---

[10] This is a simple consequence of Theorem 16.

4. Intersection of full orders

In certain cases the partial order of a ring can be represented as an intersection of full orders.

Theorem 19. (FUCHS [9].) *A partial order P of an arbitrary ring R is the intersection of full orders Q_λ ($\lambda \in \Lambda$) if, and only if, the fact that*

$$H(P, -a, \varepsilon_1 a_1, \ldots, \varepsilon_n a_n)$$

is not conic for any choice of the signs $\varepsilon_i = \pm 1$ implies $a \in P$.

Assume that P is an intersection of full orders Q_λ and $a \notin P$. Then $a \notin Q_\lambda$ for some Q_λ, hence $-a \in Q_\lambda$ and $H(P, -a)$ can be extended to a full order. Theorem 1 implies that for an appropriate choice of signs ε_i, $H(P, -a, \varepsilon_1 a_1, \ldots, \varepsilon_n a_n)$ is conic. Conversely, if the condition concerning P is fulfilled and $a \notin P$, then $H(P, -a, \varepsilon_1 a_1, \ldots, \varepsilon_n a_n)$ is conic for every set $a_1, \ldots, a_n$ with a proper choice of signs ε_i. Applying Theorem 1 again, we infer the existence of a full order Q_λ containing $H(P, -a)$; thus $a \notin Q_\lambda$. Consequently, the intersection of all f. o. extensions Q_λ of P contains no element outside P. Q. E. D.

Theorem 19 becomes perceptibly simpler if there are no zero-divisors. First we note that if P is an intersection of linear orders, then it contains all even products relative to P; a product is called *even relative to P* if it belongs to P or becomes an even product after omitting the factors belonging to P. Furthermore, P is plainly division-closed. The converse also holds:

Corollary 20. *For a partial order P of a ring R without divisors of zero to be an intersection of full orders, the following conditions are necessary and sufficient:*

(a) *P contains all sums of even products relative to P;*

(b) *P is division-closed.*

All we need is to establish the sufficiency. We go back to Theorem 7 in order to derive from (a) and (b) that P is an intersection of full orders. Suppose that, for some $a \in R$ ($a \neq 0$), $H(P, -a)$ has no full extension. In view of Theorem 7 and (a) this means that $x_0 + x_1 = 0$ holds for some $x_0 \in P$ and for some x_1 which is a non-void sum of non-zero products whose factors are elements of P, arbitrary elements of R an even number of times and $-a$ an odd number of times. Then $x_0\, a + x_1\, a = 0$ where, clearly, $x_1(-a) \in P$ and $x_1\, a \neq 0$. Hence $x_0\, a \in P$, and (b) implies $a \in P$. An application of the preceding theorem completes the proof.

Since P is a semiring, in (a) it suffices to require that P contain all even products relative to P.

Corollary 21. *A partial order P of a field F is an intersection of linear orders exactly when P contains all (sums of) products of squares.*

The condition is trivially necessary. It is also sufficient, for in the case of fields an even product relative to P is a product of an element of P and of squares, i. e. (a) of Corollary 20 is satisfied. Further, because of $ab^{-1} = ab(b^{-1})^2$, (b) also holds, and so Corollary 20 implies the assertion.

In the commutative case the condition reduces to that P contains all (sums of) squares.

5. Vector rings

We call a p. o. ring R a *vector ring* if it is a subdirect sum of f. o. rings R_λ. This means that every element of a vector ring is a vector $a = \langle \ldots, a_\lambda, \ldots \rangle$ with a_λ in the f. o. ring R_λ, and a vector is ≥ 0 if, and only if, every a_λ is ≥ 0.

It is evident that if the partial order of R is an intersection of full orders, then R is necessarily a vector ring. For in this case one can choose for the rings R_λ the ring R itself endowed with the different linear extensions of its partial order.

The analogues of the lemma and Theorem 20 in Chapter III, 6 are valid for rings:

Lemma. *A necessary and sufficient condition that a p. o. ring R be a vector ring is the existence of a representation*

$$P = \cap\, T_\lambda$$

of its positive cone P where

1) *the T_λ are convex subsemirings containing P,*

2) $x \in R \setminus T_\lambda$ implies $-x \in T_\lambda$.

Theorem 22. *A p. o. ring R with the positive cone P is a vector ring if, and only if,*

$$\cap H(P, \varepsilon_1 a_1, \ldots, \varepsilon_n a_n) = P$$

holds for every finite set of elements $a_1, \ldots, a_n$ of R where the intersection is to be taken for all 2^n choices of signs $\varepsilon_i = \pm 1$.

The proofs run parallel to those in the case of vector groups, and may be left to the reader.[11]

---

[11] $T_\lambda \cap -T_\lambda$ is an ideal, for if $a \in T_\lambda \cap -T_\lambda$ and $b \in R$, then either $b \in T_\lambda$ or $-b \in T_\lambda$ whence $\pm ab \in T_\lambda$ in all cases.

CHAPTER VIII

FULLY ORDERED RINGS AND FIELDS

1. Archimedean fully ordered rings

A p. o. ring R is called *Archimedean* if its additive group R_+ is Archimedean as a p. o. group.

In the case of fields it is easy to recognize the Archimedean ordering. Namely, a f. o. field K is Archimedean if, and only if, to every strictly positive $a \in K$ there exists a natural integer n such that $n > a$ [or equivalently, $n^{-1} < a$]. In fact, in fields the inequalities $na > b$, $n > ba^{-1}$, $n^{-1} < ab^{-1}$ are equivalent $(a, b > 0)$.

The Archimedean f. o. rings can be completely described:

Theorem 1. (PICKERT [1], HION [1].[1]) *An Archimedean f. o. ring is either a zero-ring with additive group o-isomorphic to a subgroup of the real numbers or o-isomorphic to a uniquely determined subring of the real number field, taken with the usual ordering. Thus an Archimedean f. o. ring is always associative and commutative.*[2]

By HÖLDER's Theorem 1 (Chapter IV), the additive group R_+ of an Archimedean f. o. ring R is o-isomorphic to a subgroup of the real number field V. For convenience's sake we assume R_+ to be imbedded in V. Consider the mapping $x \to a \cdot x$ of R_+ into itself ($a, x \in R$; the dot indicates multiplication in R, while juxtaposition will denote products of real numbers). This is an o-homomorphism for $a \geq 0$; hence, by Proposition 2 of Chapter IV, there exists a real number $r_a \geq 0$ such that

$$a \cdot x = r_a x \quad \text{for all } x \in R.$$

---

[1] The commutativity of Archimedean f. o. fields has been proved by HILBERT [1]. For f. o. rings with strict ordering the theorem has been proved by PICKERT [1], the theorem as it stands is due to HION [1]. See also TALLINI [1].

[2] Let us mention here that WAGNER [1] established a commutativity theorem of another kind: a f. o. associative ring containing in its centre a subfield of the real numbers is necessarily commutative whenever it satisfies a non-trivial polynomial identity. For a simpler proof see ALBERT [2].

Putting $r_a = -r_{-a}$ for $a < 0$, it is readily seen that the correspondence

$$\varphi: a \to r_a$$

satisfies $r_{a+b} = r_a + r_b$, and so it is an o-homomorphism between the real groups R_+ and, say, $S_+ \subseteq V$. Again by the same Proposition 2, we have $sa = r_a$ for some real number $s \geq 0$ and for all $a \in R$. If $s = 0$, then R is a zero-ring and we obtain the first part of the theorem. If $s > 0$, then φ is an o-isomorphism not only in the group-theoretic sense, but ring-theoretically too, since $a \cdot b = r_a b = (sa)b = s(ab)$ implies

$$r_{a.b} = s(a \cdot b) = s[s(ab)] = (sa)(sb) = r_a r_b$$

for all $a, b \in R$. Hence the elements of S_+ form a subring S of V to which R is o-isomorphic.

In order to establish unicity, we show that if ψ is an o-isomorphism between two subrings A, B of the real field, then $A = B$ and ψ is the identity. From Proposition 2 of Chapter IV we infer the existence of a real number $r > 0$ such that $\psi(a) = ra \; (\in B)$ for all $a \in A$. Now $\psi(a_1)\psi(a_2) = \psi(a_1 a_2)$ implies $r^2 = r$ whence $r = 1$, for A, B are not zero-rings. This completes the proof of Theorem 1 as well as that of

Corollary 2.[3] *An Archimedean f. o. ring has no o-automorphism other than the identity unless it is a zero-ring.*

The Archimedean character is preserved obviously under passage to a ring of quotients. The same holds for algebraic extensions, moreover:

Proposition 3.[4] *A f. o. algebraic algebra L over an Archimedean f. o. subfield K is o-isomorphic to a subfield of the real numbers.*

Let $a \in L \; (a > 0)$ satisfy

$$f(a) = a^m + a_1 a^{m-1} + \ldots + a_m = 0 \quad (a_j \in K)$$

and choose $c \in K$ such that $c > 0$, $c \geq 1 - a_j \; (j = 1, \ldots, m)$.

---

[3] For rings with identity this has been proved by PICKERT [1].

[4] For finite extensions of K the assertion has been proved by PICKERT [1] and for semi-simple algebras over K by BANASCHEWSKI [1].

Then $a \geq c$ is impossible, since this inequality would imply $a_j \geq 1 - a$ and

$$f(a) \geq a^m + (1 - a) (a^{m-1} + \ldots + 1) = 1.$$

Hence $a < c$ for some $c \in K$, and it follows easily that L is Archimedean. By Theorem 1, L is o-isomorphic to a subring of the real numbers. This must be a field, for a^{-1} belongs to the subring generated by K and a.

2. The Archimedean classes

Let R be a p. o. ring and define the Archimedean equivalence and Archimedean classes with respect to addition as done in Chapter IV, 1.[5] There we have seen that the Archimedean classes of a p. o. group form a p. o. set. In our present case we can moreover prove:

Theorem 4.[6] *The Archimedean classes of a p. o. ring form a p. o. groupoid under complex multiplication, those of a p. o. field form a p. o. group. If the ring (the field) is f. o., so are the classes.*

Let $\varkappa(x)$ denote the Archimedean class of the element x (> 0) of the p. o. ring R. First, we verify that $\varkappa(a) \leq \varkappa(b)$ implies $\varkappa(ac) \leq \varkappa(bc)$ for every $c \in R$ ($c > 0$). We can find a natural integer n such that $a < nb$. Hence $ac \leq nbc$. If here $bc = 0$, then also $ac = 0$, while if $bc \neq 0$, then $ac < (n + 1)bc$. Thus in either case $\varkappa(ac) \leq \varkappa(bc)$, as claimed. A similar argument applies to show that $\varkappa(a) \leq \varkappa(b)$ implies $\varkappa(ca) \leq \varkappa(cb)$. Therefore $\varkappa(a) = \varkappa(b)$ and $\varkappa(a') = \varkappa(b')$ imply $\varkappa(aa') = \varkappa(bb')$; and we conclude that the complex product of two Archimedean classes, $\varkappa(a)$ and $\varkappa(a')$, belongs to a third class $\varkappa(aa')$, i. e. $\varkappa(a) \cdot \varkappa(a') = = \varkappa(aa')$. What has been proved also shows that the monotony

---

[5] If the ring happens to be an algebra over a f. o. field F, then the Archimedean classes are formed with respect to F, i. e. an element and its multiples by positive elements of the underlying field belong to the same Archimedean class.

[6] This result has been proved for f. o. fields by BIRKHOFF [3]. For the definition of p. o. groupoid see Chapter X. — In the case of fields the class of zero must be omitted.

laws hold, thus the classes form a p. o. groupoid. If R is a field, then $\varkappa(e)$ is an identity for the Archimedean classes and $\varkappa(a^{-1})$ is the inverse for $\varkappa(a)$, i. e. in this case we obtain a p. o. group.

It is natural to consider the mapping $a \to \varkappa(a)$ from the positive cone of R onto the p. o. groupoid A of Archimedean classes of R as a *natural valuation* of R. A number of properties of R are reflected in A.

Let us suppose that R is an associative f. o. ring without divisors of zero which is not Archimedean, but which has an Archimedean subring $T \neq 0$. For all $x \in R$, $a \in T$ $(x, a > 0)$ we have the relation $xa \sim x$, $ax \sim x$ (Archimedean equivalence), since e. g. from $xa \ll x$ $(x \ll xa)$ we should get $xa^2 \ll xa$ $(xa \ll xa^2)$ whence $a^2 \ll a$ $(a \ll a^2)$ contrary to the hypothesis on T. The elements $x \in R$ with $|x| \sim a$ for some $a \in T$ $(a > 0)$ form a subring T^* of R containing T. Clearly, T^* is the maximal Archimedean subring containing T. Moreover, $T^* \neq R$, thus there exists an element $x \in R$ such that either $x \gg a$ or $x \ll a$ for all $a \in T^*$. In the first alternative, $x^k \sim ax^k \ll x^{k+1}$ for all $a \in T^*$, therefore $x \ll x^2 \ll x^3 \ll \ldots$. In a polynomial

$$f(x) = a_0 + a_1 x + a_2 x^2 + \ldots + a_r x^r \quad (a_i \in T^*, a_r \neq 0)$$

the terms $a_i x^i$ (if not zero) belong to different Archimedean classes, and so $f(x) \sim x^r$. Thus $f(x) > 0$ if, and only if, $a_r > 0$. In the second alternative we have $x \gg x^2 \gg x^3 \gg \ldots$, and $f(x) > 0$ if, and only if, $a_i > 0$ where $a_0 = \ldots = a_{i-1} = 0$, $a_i \neq 0$. This proves:

Theorem 5. (BANASCHEWSKI [1].) *A f. o. domain of integrity which is not Archimedean but has an Archimedean subring $T \neq 0$ contains a maximal Archimedean subring $T^* \supseteq T$. If $x \in R \setminus T^*$, $x > 0$, then the subring generated by T^* and x is o-isomorphic to the lexicographically or antilexicographically ordered polynomial ring $T^*[x]$.*

The special case when R is a field and T is its prime field is also of interest; cf. Proposition 3.

The ordering of non-Archimedean f. o. fields was first investigated by BAER [1].

3. O-rings with divisors of zero

We have a fair amount of information on O-rings without divisors of zero, but little on O-rings possessing divisors of zero. Turning to this case, we shall throughout this section consider only associative rings, because non-associativity would greatly add to the difficulties of the argument.

The following results show that O-rings with zero-divisors are somewhat exceptional.

Lemma. *The set N_n of all elements a in an associative O-ring R which satisfy $a^n = 0$ is an ideal of R whose nth power vanishes. N_n is convex in every full order of R, and it consists of full Archimedean classes.*

For the proof we assume that R is f. o. in some way. If $a \in N_n$, $b \in R$ are positive elements and, say, $ab \geq ba$, then

$$a^2 b^2 \geq abab = (ab)^2, \ldots,$$

$$a^n b^n = a(a^{n-1} b^{n-1})b \geq a(ab)^{n-1} b \geq a(ba)^{n-1} b = (ab)^n.$$

Hence $ab \in N_n$ and similarly $ba \in N_n$. If $a_1, a_2 \in N_n$, and, say, $0 \leq a_2 \leq a_1$, then $(a_1 \pm a_2)^n \leq (2a_1)^n = 0$, thus $a_1 \pm a_2 \in N_n$. We see that N_n is an ideal of R. If $a_1, \ldots, a_n \in N_n$ and if we put $a = \max(a_1, -a_1, \ldots, a_n, -a_n)$, then $a \in N_n$ and $0 = -a^n \leq a_1 \ldots a_n \leq a^n = 0$; hence the nth power of N_n vanishes, as stated.

The convexity of N_n is obvious. If $b \sim a \in N_n$, then $b < ma$ for some natural integer m and we get $0 \leq b^n \leq (ma)^n = 0$, $b \in N_n$. This completes the proof.

Theorem 6.[7] *The nilpotent elements of an associative O-ring R form an ideal N, and the factor ring R/N contains no divisors of zero. N is a convex ideal in every full order of R.*

N is the union of the ascending chain $N_1 \subseteq \ldots \subseteq N_n \subseteq \ldots$ of ideals, thus it is itself an ideal, and it is convex in every full order of R. If $a, b \in R$ satisfy $ab \in N$ and say, $0 < a \leq b$ in some full order of R, then $0 \leq a^2 \leq ab \in N$. This shows that $a^2 \in N$, $a \in N$, i. e. R/N has no divisors of zero.

---

[7] This theorem has been proved for f. o. semigroups by HION [3].

Note that it follows that in associative O-rings the existence of divisors of zero implies that of non-zero nilpotent ideals.

Let us turn to O-algebras over fields.

Theorem 7. (ZEMMER [1].) *Let A be an associative O-algebra of finite rank over a commutative field F and suppose that A is neither nilpotent nor a field. Then A is the direct sum of three vector spaces over F :*

$$A = B + C + D.$$

One of these, B, is a commutative O-field, and if e is the unit element of B and N the radical of A, then $C = eNe$ and D is either the right or the left annihilator of e. Finally, in every full order of A the elements of D are infinitely small compared to those of C, and these are infinitely small compared to those of B.

By Proposition 6 of Chapter VII, $N \neq 0$, and from the remark made after it we conclude that A/N is a commutative O-field, hence it is separable. By WEDDERBURN's principal theorem on algebras[8] one has $A = B + N$ for some subspace $B \cong A/N$. Let $L_e = A(1 - e)$ be the set of left, and $R_e = (1 - e)A$ the set of right annihilators of the identity e of B. Then either eL_e or R_ee vanishes; for if $0 < x \in eL_e$ and $0 < y \in R_ee$ in some full order of A, and e. g. $x \geq y$, then $0 > -y = -ye = (x - y)e \geq 0$ is a contradiction. Assume, for definiteness, that $eL_e = 0$. The left-sided Peirce decomposition $A = Ae + L_e$ implies $eA = eAe$, and so the right-sided can be written in the form $A = eAe + R_e$. N being an ideal of A, we have $eAe = B + eNe$. A simple substitution establishes that $A = B + eNe + R_e$.

Since $e \notin N$, $eR_e \subseteq N$ and A/N has no divisors of zero, we have $R_e \subseteq N$, and so $N = eNe + R_e$. Because of the convexity of N, it remains to prove that $y > z$ for all strictly positive $y \in eNe$, $z \in R_e$. If we had $y \leq z$, then $0 = eze \geq eye = y$ would imply $y = 0$, a contradiction. This completes the proof.

---

[8] For the results we need from the theory of algebras see e. g. A. A. ALBERT, *Structure of algebras* (New York, 1939).

4. *o*-simple fully ordered rings

A p. o. ring R is called *o-simple* if it contains no convex ideals other than 0 and R. An Archimedean f. o. ring is, for instance, *o*-simple because of Theorem 1.

Theorem 8. (D. G. JOHNSON [1].) *Let R be a f. o. associative ring and I a minimal convex ideal of R. If $I^2 \neq 0$, then R is o-simple (and $I = R$).*

If the minimal convex ideal[9] I of R satisfies $I^2 \neq 0$, then R contains no nilpotent ideals, and therefore by Theorem 6 no divisors of zero. Thus $IaI \neq 0$ for an arbitrary $a > 0$ in I. The set J of all $d \in R$ satisfying

$$|d| \leq bac \quad \text{for some } b, c \in I$$

is a convex ideal, since $d_1, d_2 \in J$, i. e. $|d_i| \leq b_i ac_i$ $(b_i, c_i \in I)$ implies that $|d_1 \pm d_2| \leq |d_1| + |d_2| \leq 2 \max (b_i ac_i) =$ $= \max (2b_i)ac_i$ and $|xd_1| = |x||d_1| \leq (|x|b_1)ac_1$, $|d_1 x| \leq$ $\leq b_1 a(c_1|x|)$. In view of $0 \subset J \subseteq I$ and the minimality of I we obtain $J = I$. Thus a satisfies $a \leq bac$ for some $b, c \in I$. We cannot have $ab < a$ and $ba < a$ for all positive $b \in I$, for then $(ba)c \leq ac < a$ would be a contradiction. Hence $ab \geq a$ or $ba \geq a$ for some positive $b \in I$. In the first case $abx \geq ax$ for all positive $x \in R$. Hence $bx \geq x$ since R contains no divisors of zero, and therefore $I = R$, i. e. R is *o*-simple. The same conclusion holds in the second case. Q. E. D.

The next result is somewhat surprising.

Theorem 9. (D. G. JOHNSON [1].) *An o-simple f. o. associative ring R contains no non-trivial convex one-sided ideals.*

If R contains zero-divisors, then (in the notation of the preceding section) $N_2 \neq 0$ whence $N_2 = R$ and R is a zero-ring. In this case R is commutative and the assertion is obvious. Consequently, it suffices to verify that every f. o. associative ring S without divisors of zero that contains a non-trivial

---

[9] Clearly, there may exist only one minimal convex ideal. Its existence is equivalent to the fact that the f. o. ring is subdirectly irreducible. (Example 3 in Chapter VI, **2** with the antilexicographic full order is a subdirectly reducible f. o. ring.)

convex left ideal L contains a non-trivial convex ideal. If S has an identity e, then clearly $e \notin L$, furthermore the convex ideal I generated by L does not contain e. For if, on the contrary,

$$e \leq a_1 x_1 + \ldots + a_n x_n \quad (0 < a_i \in L, \ 0 < x_i \in S),$$

then also

$$e \leq ax \quad (a = a_1 + \ldots + a_n, \ x = \max x_i),$$

whence $a \leq axa$. But every element of L is less than e, hence $xa < e$ and $axa < a$, a contradiction. If S has no identity, then by Corollary 7 of Chapter VI we can embed S in a f. o. ring S^* with identity e (with preservation of ordering) such that neither S^* contains zero-divisors. L is a left ideal in S^*, and the convex left ideal L^* generated by L in S^* does not contain e. For if we had $e < a$ for some $a \in L$, then by the definition of full order in S^* we should get $b < ab$ for a certain positive $b \in S$ whence $xb < xab$ and $x < xa$ ($\in L$) for all strictly positive $x \in S$, i. e. $L = S$. Thus L^* is a non-trivial convex left ideal of S^*, and by what has been proved we conclude that the convex ideal I^* of S^* generated by L^* differs from S^*. The intersection $I = I^* \cap S$ is a convex ideal of S which cannot coincide with S. This is trivial if L is already an ideal, for then $I = L$. If this is not the case, then $by \notin L$ for some positive elements $b \in L$, $y \in S$, whence $b < by$ and $e < y$ in S^*, thus $y \notin I$. This completes the proof.

A simple consequence of the theorem is that every maximal convex ideal in a f. o. associative ring is at the same time a maximal convex left (and right) ideal.

In general, a f. o. associative ring may contain convex one-sided ideals that are not two-sided. Examples have been given by HOLLAND [1] and D. G. JOHNSON [1]. JOHNSON exhibits a f. o. ring A, namely the semigroup ring of the free semigroup without identity generated by two elements, over the ring of integers, (under a suitable definition of full order) such that every f. o. associative ring without divisors of zero containing a convex left ideal that is not two-sided, contains a subring o-isomorphic to A.

5. Formal power series fields

The well-known construction of formal power series over a field can be generalized in our situation as follows.

Let G be a f. o. group and F a field. We define "formal infinite sums"

$$(1) \qquad \Phi = \sum_{a \in G} \Phi_a \, a \quad (\Phi_a \in F)$$

with the proviso that the support

$$S(\Phi) = [a \in G \mid \Phi_a \neq 0]$$

is well-ordered in the full order of G. Then (1) will be called a *formal power series on G over F*. The set of all these formal power series is denoted by $F[[G]]$.

If Φ and Ψ are two formal power series, then their sum $\Phi + \Psi$ is defined in the obvious manner by

$$(\Phi + \Psi)_a = \Phi_a + \Psi_a.$$

Thus $\Phi + \Psi$ is again a formal power series and so is $\lambda \Phi$ for every $\lambda \in F$ where

$$(\lambda \Phi)_a = \lambda \Phi_a.$$

Thus $F[[G]]$ is a left vector space over F. We turn it into a ring by defining a product, depending on a certain choice of automorphisms of F, as follows. Let ω be an antihomomorphism of G into the automorphism group of F; thus $\omega(b)$ is, for every $b \in G$, an automorphism of F, and

$$\omega(bc) = \omega(c) \, \omega(b) \qquad \text{for all } b, c \in G.$$

Then we define the product $\Phi\Psi$ of Φ and Ψ by[10]

$$(2) \qquad (\Phi\Psi)_a = \sum_{bc=a} \Phi_b \, \Psi_c^{\omega(b)}.$$

Note that

---

[10] One could instead use homomorphisms ω^* of G into the automorphism group of F and define

$$(\Phi\Psi)_a = \sum_{bc=a} \Phi_b^{\omega^*(c)} \Psi_c.$$

A) The sum in (2) is finite: only a finite number of b's exist such that $\Phi_b \neq 0$ and $\Psi_{b^{-1}a} \neq 0$. Indeed, there exists no infinite decreasing sequence of b's with $\Phi_b \neq 0$ and there exists no infinite increasing sequence of b's (i. e. decreasing sequence of $b^{-1}a$'s) with $\Psi_{b^{-1}a} \neq 0$.

B) The associative and distributive laws are fulfilled because of the assumptions on the ω's. Thus $F[[G]]$ is a ring; it will be denoted by $F_\omega[[G]]$.

C) For all $b \in G$ we have

$$\Phi b = \sum \Phi_a(ab) \quad \text{and} \quad b\Phi = \sum \Phi_a^{\omega(b)}(ba).$$

Next we prove the following important

Lemma. *Let* $\Phi \in F_\omega[[G]]$ *with* $\Phi_a \neq 0$ *only for* $a > e$. *Then for all* $\lambda_n \in F$ *the infinite sum*

$$\sum_{n=1}^{\infty}{}' \lambda_n \Phi^n$$

is meaningful.

It suffices to verify that 1) the union of the supports $S(\Phi^n)$ $(n = 1, 2, \ldots)$ is well-ordered in the full order of G, and 2) for each fixed $a \in G$, there exist only a finite number of integers n such that $(\Phi^n)_a \neq 0$.

Evidently, assertion 1) will follow at once if we shall have shown that there does not exist any infinite sequence

(3)
$$u_1 = a_{11} \ldots a_{1n_1} > \ldots > u_i = a_{i1} \ldots a_{in_i} > \ldots$$
$$\text{with } \Phi_{a_{ik}} \neq 0 \ (a_{ik} \in G).$$

By way of contradiction, assume the existence of a sequence (3). Note that in the f. o. group G we have obviously

$$\{u_i\}_\square = \{\max_k a_{ik}\}_\square$$

and that by the definition of formal power series in every subset of the $\{a\}_\square$ with $\Phi_a \neq 0$ there exists a smallest subgroup; thus among the $\{u_i\}_\square$ there must exist a smallest subgroup U of G. We may suppose that the sequence (3) is chosen so that this smallest subgroup is as small as possible. In view of $\{u_1\}_\square \supseteq \ldots \supseteq \{u_i\}_\square \supseteq \ldots$, we may moreover assume without loss of generality that

$$\{u_i\}_\square = U \quad (i = 1, 2, \ldots).$$

Now pick out for every i some $a_i^* = a_{ik_i}$ with $\{a_{ik_i}\}_\square = \{u_i\}_\square = U$. There may exist several $a \in G$ satisfying $\Phi_a \neq 0$ and $\{a\}_\square = U$, but among them there certainly exists one, say a^*, which is the smallest in the full order of G. For this a^* we have $a^* \leq a_1^* \leq u_1$ where $\{a^*\}_\square = \{a_1^*\}_\square = \{u_1\}_\square$. From Chapter IV, **3** it follows that $u_1 \leq a^{*r}$ for some natural integer r, and hence

$$u_i \leq a^{*r} \quad (i = 1, 2, \ldots).$$

Every u_i is of one of the forms

$$u_i = a_i^*, \quad u_i = v_i a_i^*, \quad u_i = a_i^* w_i, \quad u_i = v_i a_i^* w_i$$

where v_i, w_i denote certain products of the a_{ik}. As among the a_i^* there cannot exist any properly decreasing infinite sequence, only a finite number of the u_i can be of the first form. Thus either among the v_i or among the w_i there exists a properly decreasing sequence; for definiteness suppose that the first alternative occurs: $v_{i_1} > \ldots > v_{i_j} > \ldots$. This is again a sequence of the form (3) where we have $\{v_{i_j}\}_\square = U$ (because $v_i \leq u_i$) and $v_i \leq a^{*r-1}$ $(i = 1, 2, \ldots)$ (because $a^* \leq a_i^*$ and $u_i \leq a^{*r}$). Thus from the u_i we have constructed another sequence v_{i_j} with the same minimal property, and in addition with a smaller r. This leads to an obvious contradiction which establishes 1).

For the proof of 2), assume the existence of an $a \in G$ such that

(4)
$$a = a_{i1} \ldots a_{in_i} \quad (i = 1, 2, \ldots)$$
$$\text{with} \quad n_1 < \ldots < n_i < \ldots \quad \text{and} \quad \Phi_{a_{ik}} \neq 0$$

and select the smallest a of this type (this can be done in view of 1)). The sequence $a_{11}, \ldots, a_{i1}, \ldots$ has a non-decreasing infinite subsequence which may be denoted by the same symbols, $a_{11} \leq \ldots \leq a_{i1} \leq \ldots$. The sequence $a'_i = a_{i2} \ldots a_{in_i}$ $(i = 1, 2, \ldots)$ is not increasing. $a'_1 \geq \ldots \geq a'_i \geq \ldots$, and what has been proved in the preceding paragraph shows that the a'_i are equal from some index j on, $a'_j = a'_{j+1} = \ldots$. But $a'_j < a$ which contradicts the choice of a in (4). This completes the proof of 2), and hence that of the lemma.

Consider the set Γ of all $e + \Phi$ with $\Phi \in F_\omega[[G]]$ such that $\Phi_a \neq 0$ only for $a > e$. This Γ is a group under the composition

$$(e + \Phi)(e + \Psi) = e + (\Phi + \Psi + \Phi\Psi).$$

In fact, $e + \sum_{n=1}^{\infty} (-\Phi)^n$ is the inverse of $e + \Phi$; this exists because of the lemma.

Now we are ready to prove

Theorem 10. (HAHN [1], NEUMANN [2].)[11] *The formal power series on a f. o. group G over a field F form a field $F_\omega[[G]]$.*

We know already that $F_\omega[[G]]$ is a ring. An element $\Psi \neq 0$ of $F_\omega[[G]]$ may be written in the form $\Psi = \lambda b(e + \Phi)$ with $\lambda \in F$, $b \in G$, $e + \Phi \in \Gamma$. If $e + \overline{\Phi}$ is the inverse of $e + \Phi$ in Γ, then $X = (e + \overline{\Phi}) b^{-1} \lambda^{-1} \in F_\omega[[G]]$ satisfies $\Psi X = X \Psi = = e$. Clearly, e acts as the identity element of $F_\omega[[G]]$.

We proceed to the case when the coefficient field is f. o. too. In order to ensure that products of positive elements are again positive, we impose a further restriction on the ω's:

Corollary 11. (HAHN [1], NEUMANN [2].) *If G, F are as in the preceding theorem and F is f. o., then $F_\omega[[G]]$ can be f. o. (in a natural way) provided the ω's are o-automorphisms.*

Define $\Psi = \lambda b (e + \Phi)$ $(e + \Phi \in \Gamma)$ positive if $\lambda > 0$. Then it is routine to check the requisite postulates for the full order.

If we take for the ω's the identity automorphism ι, the existence of $F_\iota[[G]]$ is ensured and we get

Corollary 12. (HILBERT [1].) *Every f. o. group can be embedded in the multiplicative group of a f. o. field.*

---

[11] The commutative case is due to HAHN, while the first complete proof in the non-commutative case has been given by NEUMANN. The ω's have been defined above.

The group ring of G with coefficients in F is contained in $F_\iota[[G]]$, therefore

Corollary 13. (MAL'CEV [1], NEUMANN [2].)[12] *The group ring of a f. o. group G over a (f. o.) field F can be embedded in a (f. o.) field.*

NEUMANN [2] considers power series fields more generally. He takes factor sets $\gamma(a, b) \in F$ subject to

$$\lambda^{\omega(b)\omega(a)} \gamma(a, b) = \gamma(a, b) \lambda^{\omega(ab)},$$

$$\gamma(ab, c) \gamma(a, b)^{\omega(c)} = \gamma(a, bc) \gamma(b, c)$$

for all $a, b, c \in G$, $\lambda \in F$.[13]

For two elements, Φ and Ψ, of a f. o. field $F_\omega[[G]]$ we have $\Phi \ll \Psi$ if, and only if, the first $a \in G$ with $\Phi_a \neq 0$ is less than the first $b \in G$ with $\Psi_b \neq 0$. This is an obvious consequence of the definition of ordering in $F_\omega[[G]]$. Hence Φ and Ψ are Archimedean equivalent with respect to F if, and only if, $a = b$. We conclude that the Archimedean classes of $F_\omega[[G]]$ with respect to F are in a one-to-one correspondence with the elements of G:

Proposition 14. *The f. o. group of the Archimedean classes of the formal power series field $F_\omega[[G]]$ with respect to the f. o. field F is in the natural way o-isomorphic to G.*

Let us note that every f. o. commutative field F can be embedded in the field $R[[G]]$ where R is the real field, G is the f. o. group of Archimedean classes of F and ω is the map of G onto the identity automorphism of R.

6. Completion of fully ordered fields

If we come to the problem of embedding p. o. fields in complete l. o. fields, and want to apply the Dedekind or the Cantor process to the additive group of the field, then the result is only exceptionally a field again. This situation may be partly elucidated by the following lemma.

---

[12] A special case of this result is due to MOUFANG [1].
[13] A further generalization to non-associative systems has been considered by ZELINSKY [2].

Lemma. *A l. o. commutative field K with division-closed positive cone is necessarily f. o.*

Let $a \in K$, $a \neq 0$, and define $b = (a^+)^2 \, |a|^{-1}$. By hypothesis, $|a|^2 \, |a|^{-1} = |a| > 0$ implies $|a|^{-1} > 0$, and so $b \geq 0$. From $|a| = a^+ - a^-$, $a = a^+ + a^-$ we obtain that $b - a = (a^-)^2 \, |a|^{-1} \geq 0$ and $a^+ - b = -a^- \, a^+ \, |a|^{-1} \geq 0$. This shows that $b = a^+$ and hence the last inequality implies $a^- \, a^+ = 0$, i. e. either $a > 0$ or $a < 0$.

Accordingly, a directed commutative field in the important division-closed case cannot be embedded in a complete l. o. field with preservation of ordering unless it is f. o. Another kind of unsurmountable difficulty arises in the non-commutative case, so that it is rather natural to restrict ourselves *ab initio* to f. o. commutative fields.

Let F be a f. o. commutative field. We adapt the Cantor process to obtain a field enjoying certain completeness properties.

Define fundamental sequences and null sequences of elements of F in the usual way. The fundamental sequences form a ring R with identity in which the null sequences form a maximal ideal N (the proof of this, which follows classical lines, can be supplied by the reader).[14] The factor ring $R/N = = F^*$ is therefore a field which may be f. o. in a natural way so that the canonical map $a \to (a, \ldots, a, \ldots)$ from F into R followed by the natural homomorphism of R onto F^* is an o-isomorphism between F and a subfield of F^*. Identifying F with this subfield, it follows that any element of F^* is the limit of a fundamental sequence in F and every fundamental sequence in F has a limit in F^*.

It is to be emphasized that the closure F^* of F has not in general the property that every bounded set in F has a l. u. b. and a g. l. b. in F^*. This happens if F is Archimedean, but in no other case—as is easily seen. Let us formulate the familiar theorem:

---

[14] We refer to DUBREIL [1]; his method seems to be most adequate to show how to use the Cantor process in the general case.

Theorem 15. *If F is an Archimedean f. o. field, then the f. o. ring $F^* = R/N$ obtained from F by the Cantor process is a complete f. o. field. Hence it is o-isomorphic to the real number field.*

If commutativity is removed, then the process sketched does not lead in general to a field F^*. It has been shown by NEUMANN [2] that every f. o. field K can be embedded in a f. o. field $K(F^*)$ continuing the ordering of K and containing in its centre the real number field F^*. This result has been generalized by JAEGER [1] who proved that the same is true if F is an arbitrary subfield of the centre of K and F^* is the field obtained from F by the Cantor process.

CHAPTER IX

LATTICE-ORDERED RINGS

1. General properties of lattice-ordered rings

If the additive group of a p. o. ring R is l. o., we call R a *l. o. ring*.[1] The zero-ring on an additive Abelian l. o. group and every f. o. ring furnish trivial examples of l. o. rings. Less trivial and more important examples of l. o. rings are examples 5, 6 and 7 in Chapter VI, **2**.

The results established in Chapter V for l. o. groups are automatically valid in l. o. rings R. For convenience' sake we list here the most important ones, together with some simple remarks on the multiplication in the ring.

A) For all $a, b, c \in R$ we have

$$(a \lor b) + c = (a + c) \lor (b + c), \quad (a \land b) + c = (a + c) \land (b + c),$$

$$-(a \lor b) = -a \land -b, \quad -(a \land b) = -a \lor -b,$$

$$(a \lor b) + (a \land b) = a + b.$$

B) If $c \geq 0$, then for all $a, b \in R$

$$(a \lor b)c \geq ac \lor bc, \quad c(a \lor b) \geq ca \lor cb,$$

$$(a \land b)c \leq ac \land bc, \quad c(a \land b) \leq ca \land cb.$$

These inequalities follow from the fact that inequalities like $a \lor b \geq a$ or $\geq b$ can be multiplied by a positive element.

C) The positive and negative parts of a and the absolute of a are defined now as

$$a^+ = a \lor 0, \quad a^- = a \land 0, \quad |a| = a \lor -a,$$

and they are connected by the equations

$$a = a^+ + a^-, \quad |a| = a^+ - a^-.$$

---

[1] The general theory of l. o. rings is due to BIRKHOFF—PIERCE [1].

We have further

$$a^+ \wedge -a^- = 0, \quad |a+b| \leq |a| + |b|$$

for all $a, b \in R$.

D) We have

$$|ab| \leq |a||b|,$$

for $ab = a^+ b^+ + a^+ b^- + a^- b^+ + a^- b^- \leq a^+ b^+ - a^+ b^- - a^- b^+ + a^- b^- = |a||b|$ and similarly $-|a||b| \leq ab$.

E) Define an *L-ideal* of a l. o. ring R as a subset I which is (i) an ideal of the ring R in the algebraic sense, and (ii) a convex sublattice of R. Just as in the case of l. o. groups, one verifies easily that (ii) may be replaced by condition (iii) $a \in I$, $x \in R$ and $|x| \leq |a|$ imply $x \in I$. The results on l-ideals of l. o. groups extend readily to L-ideals:

(a) the partitions defined by the congruence relations of a l. o. ring R are just the partitions of R into the cosets of its distinct L-ideals;

(b) the sum $I + J$ (or g. c. d.) of two L-ideals I and J is again an L-ideal; it consists of all sums $a + b$ with $a \in I$, $b \in J$;

(c) the L-ideals of a l. o. ring R form a complete (and distributive) sublattice of the lattice of all l-ideals of R_+;

(d) the factor ring R/I of a l. o. ring R with respect to an L-ideal I is again a l. o. ring under the induced ordering relation, and there is a natural one-to-one correspondence between the L-ideals of R/I and those of R containing I.

F) The product of two L-ideals, defined as usual in ring theory, will not in general be an L-ideal again. In order to remedy this deficiency, we introduce the *L-product* $I \cdot J$ of two L-ideals I and J as the set of all $x \in R$ with $|x| \leq \leq \sum |a_i||b_i|$ for suitable $a_i \in I, b_i \in J$. Because of $\sum |a_i||b_i| \leq \leq \sum |a_i| \cdot \sum |b_i|$, this is simply the set of all $x \in R$ with

$$|x| \leq |a||b| \text{ for some } a \in I, b \in J.$$

In the associative case $I \cdot J$ is nothing else than the L-ideal generated by the ring-theoretic product of I and J. It follows that I^2 consists of all $x \in R$ such that $|x| \leq |a|^2$ for some $a \in I$ [$|a||b|$ can be replaced by $(|a| + |b|)^2$]. More generally, I^n is the totality of all elements x of R such that $|x| \leq |a|^n$ holds for a suitable $a \in I$.[2] It is readily checked that

$$I \cdot J \subseteq I \cap J,$$

$$I \cdot (J + K) = I \cdot J + I \cdot K, \quad (J + K) \cdot I = J \cdot I + K \cdot I$$

for all L-ideals I, J, K of R.

G) BIRKHOFF and PIERCE [1] gave an enumeration of all two-dimensional l. o. algebras over the real field. We wish to underline a rather elementary but important fact they prove: *the complex number field, regarded as an algebra over the real field, admits no lattice-order.* In fact, the subgroup generated by the positive cone P must exhaust all complex numbers, hence P contains two complex numbers, z and w, not collinear with 0. Then P contains all $\lambda z + \mu w$ with positive real numbers λ, μ, so that P is a sector in the complex plane with an angle $\alpha > 0$. P^2 must have an angle 2α, in contradiction to $P^2 \subseteq P$.

2. Function rings

It seems to be difficult to say much about the structure of l. o. rings. There is one case, however, in which one can give a reduction to f. o. rings. These rings are exact analogues of l. o. vector groups considered in Chapter V. Even the results bear some resemblance to those on l. o. vector groups.

The complete direct sum

$$\bar{R} = \sum_{\lambda \in \Lambda}^* R_\lambda$$

of f. o. rings R_λ is a l. o. ring where an element $\langle \ldots, a_\lambda, \ldots \rangle$ $\in \bar{R}$ is positive if, and only if, $a_\lambda \geq 0$ for every $\lambda \in \Lambda$. A *function ring*, or briefly, an *F-ring* is a ring R which is both a subring and a sublattice of such an $\bar{R}$.

An F-ring R has the following noteworthy properties which are obvious in f. o. rings and are preserved under forming a subring—sublattice of a complete direct sum:

---

[2] If associativity is not assumed, we can define I^n as the nth left power of I, $I^n = I(I(I(\ldots I)))$, and if a similar meaning is given to $|a|^n$, our assertion remains valid.

(a) if $c \geq 0$, then for all $a, b \in R$

$$(a \vee b)c = ac \vee bc, \quad c(a \vee b) = ca \vee cb,$$

$$(a \wedge b)c = ac \wedge bc, \quad c(a \wedge b) = ca \wedge cb;$$

(b) $|ab| = |a||b|$ for all $a, b \in R$;

(c) $a^2 \geq 0$ for every $a \in R$;

(d) $a \wedge b = 0$ implies $ab = 0$.

The last property shows that *an F-ring without divisors of zero must be f. o.*

The main result on F-rings is

Theorem 1.[3] *The following conditions on a l. o. ring R are equivalent:*

(i) *R is an F-ring;*

(ii) *$a \wedge b = 0$ and $c \geq 0$ imply $ca \wedge b = 0$ and $ac \wedge b = 0$;*

(iii) *for every subset X of R, the set X^* of all elements of R orthogonal to X is an L-ideal of R.*

(i) implies (ii), for (ii) trivially holds in f. o. rings.

Assume (ii), and observe that X^* is, by Proposition 12 in Chapter V, an l-ideal of the additive group R_+. Thus what we have to prove amounts to this: if $a \in X^*$ and $c \in R$, then $ac, ca \in X^*$. For positive elements this is nothing else than (ii), while for arbitrary elements it follows at once from the inequality $|ac| \leq |a||c|$. Thus (ii) implies (iii).

Since (iii) contains (ii) as a special case (when X consists of a single element), it suffices to verify that (ii) and (iii) together imply (i). We argue in much the same way as in the proof of Theorem 15 in Chapter V. We begin with the observation that (ii) may be written in the form

$$\left. \begin{array}{l} (y \vee 0)(x \vee 0) \wedge (-x \vee 0) = 0 \\ (x \vee 0)(y \vee 0) \wedge (-x \vee 0) = 0 \end{array} \right\} \quad \text{for all } x, y \in R.$$

In fact, elements $a, b \in R$ with $a \wedge b = 0$ may be represented in the form $a = x \vee 0$, $b = -x \vee 0$ for some $x \in R$. Hence

---

[3] BIRKHOFF and PIERCE proved the equivalence of (i) and (ii) in their paper [1]. They defined F-rings by condition (ii).

the class of l. o. rings satisfying (ii) is equationally definable, and thus they are subdirect sums of subdirectly irreducible l. o. rings with property (ii). It is then sufficient to show that a subdirectly irreducible l. o. ring R satisfying (iii) is f. o. If $a, b \in R$ satisfy $a \wedge b = 0$, then the set A of all $x \in R$ orthogonal to b and the set B of all $y \in R$ orthogonal to A are L-ideals of R with 0 intersection. By subdirect irreducibility, one of A and B must vanish, i. e. either $a = 0$ or $b = 0$. The proof is completed.

Corollary 2. (BIRKHOFF—PIERCE [1].) *The F-rings are equationally definable.*

In the Archimedean case one obtains:

Theorem 3. (BIRKHOFF—PIERCE [1].) *An Archimedean F-ring is associative and commutative.*

Commutativity will follow at once if we can show that in an F-ring R

$$n \left| ab - ba \right| \leq a^2 + b^2$$

holds for all natural integers n and for all $a, b \in R$. It suffices to prove this for f. o. rings and for the case $0 < b \leq a$. Then for some integer m, $nb = ma + c$ where $0 \leq c < a$, and so

$$n \left| ab - ba \right| = \left| nab - nba \right| = \left| ma^2 + ac - ma^2 - ca \right| =$$
$$= \left| ac - ca \right| \leq a^2 \leq a^2 + b^2.$$

To prove associativity, let $a, b, c \in R$ be arbitrary elements and n a natural integer. We show that in a commutative F-ring[4]

$$n \left| (ab)c - a(bc) \right| \leq 2(\left| a \right|^3 + \left| b \right|^3 + \left| c \right|^3).$$

Restricting ourselves to f. o. rings and to the case $0 < c \leq$ $\leq b \leq a$, we determine integers m, k such that $nb = ma + a_1$, $mc = ka + a_2$ where $0 \leq a_i < a$. Then

$$n \left| (ab)c - a(bc) \right| = \left| m[a^2c - a(ac)] + (aa_1)c - a(a_1 c) \right| =$$
$$= \left| ka^2 a + a^2 a_2 - kaa^2 - a(aa_2) + (aa_1)c - a(a_1 c) \right|.$$

---

[4] Observe that by commutativity, cubes have a unique meaning.

Here the terms containing k may be omitted because of commutativity. Therefore

$$n\,\big|(ab)c - a(bc)\big| \leq \big|a^2 a_2 - a(aa_2)\big| + \big|(aa_1)c - a(a_1 c)\big| \leq$$
$$\leq a^3 + a^3 \leq 2(a^3 + b^3 + c^3),$$

and the statement is proved.

Let us turn to prime L-ideals of F-rings. Recall that in an
abstract ring R an ideal P is called prime if $AB \subseteq P$ for ideals
A, B of R implies $A \subseteq P$ or $B \subseteq P$, and completely prime if
$ab \in P$ for elements a, b of R implies $a \in P$ or $b \in P$, and these
concepts are in general different. They are identical in associative and commutative rings, and the same situation holds
in associative F-rings for the *prime L-ideals*, that is, for L-
ideals that are prime ring ideals; moreover we have:

Theorem 4. (D. G. JOHNSON [1].) *For an L-ideal P of an
associative F-ring R the following conditions are equivalent*:

(i) *if $a, b \in R$ and $ab \in P$, then $a \in P$ or $b \in P$*;

(ii) *if A, B are ideals of R and $AB \subseteq P$, then $A \subseteq P$ or
$B \subseteq P$*;

(iii) *if I, J are L-ideals of R and $I \cdot J \subseteq P$, then $I \subseteq P$ or
$J \subseteq P$*;

(iv) *R/P is a f. o. ring without divisors of zero.*

It is trivial that (i) implies (ii), (ii) implies (iii) and (iv)
implies (i). Consequently, it suffices to prove that (iv) follows
from (iii). Suppose that (iii) holds, and assume, on the contrary,
that R/P is not f. o. Then R/P would be a subdirectly reducible
F-ring, and therefore there would exist two L-ideals I, J of R
such that $I \cap J = P$ and $I \supset P$, $J \supset P$. But $I \cdot J \subseteq I \cap J$
would then imply $I \subseteq P$ or $J \subseteq P$, a contradiction; that is,
R/P is f. o. Because of (iii), the f. o. ring R/P cannot contain
nilpotent L-ideals ($=$ nilpotent convex ideals), therefore in
view of Theorem 6 in Chapter VIII, R/P contains no divisors of
zero. Q. E. D.

It follows that the complement $R \setminus P$ of a prime L-ideal P
of an F-ring R is a multiplicative subsemigroup of R (or
empty). The following result is the exact analogue of a well-
known theorem (of KRULL) in commutative ideal theory:

Proposition 5. (D. G. JOHNSON [1].) *Let I be an L-ideal of an associative F-ring R and M a multiplicative subsemigroup in R such that $I \cap M = \varnothing$. Then there exists a prime L-ideal P of R that contains I and satisfies $P \cap M = \varnothing$.*

By ZORN's lemma, hypothesis implies that there exists an L-ideal P of R that contains I and is maximal with respect to the property of being disjoint from M. If J, K are L-ideals of R satisfying $J \cdot K \subseteq P$, then also $(J + P) \cdot (K + P) \subseteq P$. Here at least one of $J + P$, $K + P$ is disjoint from M, for otherwise P would contain an element of M. From the maximality of P we obtain that $J + P \subseteq P$ or $K + P \subseteq P$, that is to say, $J \subseteq P$ or $K \subseteq P$. Thus P is a prime L-ideal, indeed.

The F-rings for which the f. o. rings R_λ may be chosen so as to have certain additional properties may be described as l. o. rings with zero radical where the radical is taken in a suitable sense depending on the given property. The radicals in associative F-rings have been studied by PIERCE in his paper [2] where a general method of defining radicals is developed and by D. G. JOHNSON [1] who has given a structure theory for F-rings based on an F-ring analogue of the Jacobson radical for abstract rings. Here we confine ourselves to mentioning the following important result.

Theorem 6. (PIERCE [2].) *A necessary and sufficient condition that an associative F-ring R be o-isomorphic to a subdirect sum of f. o. rings without divisors of zero is that R contain no nilpotent elements $\neq 0$.*

If R is a subdirect sum of the stated kind, then it contains obviously no nilpotent elements $\neq 0$. Conversely, assume that the F-ring R has no nilpotent elements $\neq 0$, i. e. $x \neq 0$ implies that the semigroup $M = [x^n \mid n = 1, 2, \ldots]$ does not contain 0. By Proposition 5 there exists a prime L-ideal P_x containing no power of x. Since the intersection of all P_x with x running over all elements $\neq 0$ in R vanishes, R is o-isomorphic to a subdirect sum of the F-rings R/P_x which are by Theorem 4 f. o. rings without divisors of zero.

The last three results show that the theory of L-ideals in associative F-rings seems to have, to a certain extent, resemblance to the ideal theory in commutative and associative rings.

A special kind of F-rings, namely those in which $ab \wedge c = 0$ is equivalent to $a \wedge b \wedge c = 0$ (for all positive a, b, $c \in R$) has been discussed by GOFFMAN [3].

3. The L-radical of lattice-ordered rings

In pure ring theory several types of radical are defined. We wish to consider here the analogue of the radical for l. o. rings based on the concept of nilpotency. Here again we have to restrict ourselves to the associative case.

We begin by defining an element a of a l. o. ring R to be *strongly nilpotent* if there exists a natural integer n such that

$$(*) \qquad x_0 \cdot |a| \cdot x_1 \cdot |a| \cdot x_2 \ldots x_{n-1} \cdot |a| \cdot x_n = 0 \qquad \text{for all } x_i \in R.$$

If R is commutative, then the strong nilpotency of a is equivalent to the nilpotency of $|a|$, for $(*)$ implies $|a|^{2n+1} = 0$, while $|a|^n = 0$ implies the validity of $(*)$.

Naturally, we say that an L-ideal I of R is *nilpotent* if $I^n = 0$ for some positive integer n. By a remark in **1** F), $I^n = 0$ is equivalent to the fact that $a^n = 0$ for all a in the L-ideal I.

Lemma. *In an associative l. o. ring R, an element is strongly nilpotent if, and only if, it is contained in a nilpotent L-ideal of R.*

If I is an L-ideal satisfying $I^n = 0$, then clearly $(*)$ holds for every a in I. Conversely, if $a \in R$ satisfies $(*)$, then

$$(R \cdot |a|)^{n+1} = 0, \quad (|a| \cdot R)^{n+1} = 0 \quad \text{and} \quad (R \cdot |a| \cdot R)^n = 0.$$

Thus the $(2n + 1)$th power of a sum of elements of the form $k \cdot |a|$, $r_1 \cdot |a|$, $|a| \cdot r_2$, $r_1 \cdot |a| \cdot r_2$ vanishes (k a rational integer, r_1, $r_2 \in R$), i. e. the L-ideal generated by a is nilpotent.

An immediate consequence is

Theorem 7. (BIRKHOFF—PIERCE [1].) *In an associative l. o. ring R, the set of all strongly nilpotent elements coincides with the union of all nilpotent L-ideals. Thus it is an L-ideal N of R.*

This N will be called the *L-radical* of R. In general, it is not nilpotent. Its nilpotency can be proved if one of the chain conditions is assumed:

Theorem 8. (BIRKHOFF—PIERCE [1].) *If R is an associative l. o. ring with maximum or minimum condition on L-ideals, then the L-radical N of R is nilpotent.*

If the maximum condition holds, then there exists a maximal nilpotent *L*-ideal M, say, $M^k = 0$ for some natural integer k. If I is a nilpotent *L*-ideal, $I^n = 0$, then $(M + I)^{k+n} = 0$, thus $M + I$ is again a nilpotent *L*-ideal. Since $M + I$ cannot properly contain M, we have $I \subseteq M$, and so M is the *L*-radical of R.

Next, assume the minimum condition on *L*-ideals. There exists a natural number k such that $N^k = N^{k+1}$ where N is the *L*-radical of R. Then $M = N^k$ satisfies $M^2 = M$. Assume $M \neq 0$. We can pick out an *L*-ideal K which is minimal with respect to the properties $K \subseteq M$ and $MKM \neq 0$, as well as an element $a \in K$ such that $a > 0$ and $MaM \neq 0$. The *L*-ideal J generated by the set MaM is contained in K and satisfies $MJM \neq 0$ (for $M^2 aM^2 \neq 0$), so $J = K$. We infer that $a \in J$, i. e. a satisfies an inequality $a \leq \sum |y_i| \cdot a \cdot |z_i|$ with y_i, $z_i \in M$, or, more simply, $a \leq yaz$ with $y = \sum |y_i|$, $z = \sum |z_i| \in M$. But then

$$0 < a \leq yaz \leq y^2 az^2 \leq \ldots \leq y^m az^m \leq \ldots.$$

Here $y^m az^m = 0$ for a sufficiently large m, because $y, z \in M$ are nilpotent. This contradiction proves that $M = 0$, i. e. N is nilpotent.

So far we have been considering l. o. rings in general. We now confine our attention to *F*-rings.

Theorem 9. (BIRKHOFF—PIERCE [1], D. G. JOHNSON [1].) *In an associative F-ring R the set N_n of all elements $a \in R$ satisfying $a^n = 0$ for some fixed n is an L-ideal. The L-radical N of R consists of all nilpotent elements. N is the intersection of all prime L-ideals of R.*

If R is represented as a subdirect sum of f. o. rings, it is obvious that the nth power of a vanishes if, and only if, the nth powers of its components vanish. For f. o. rings the first two assertions have been proved in Chapter VIII, **3**, whence they hold for F-rings too. The last statement can be verified by the argument of Theorem 6.

The following analogue of WEDDERBURN—ARTIN's classical structure theorem holds.

Theorem 10. (BIRKHOFF—PIERCE [1].) *An associative F-ring with zero L-radical and with minimum condition on L-ideals is o-isomorphic to a direct sum of a finite number of o-simple f. o. rings (that are not nilpotent).*

By the preceding result, the intersection of all prime L-ideals of an associative F-ring R with zero L-radical is zero. By the minimum condition there exist a finite number of prime L-ideals, $P_1, \ldots, P_n$, such that $P_1 \cap \ldots \cap P_n = 0$. Hence R is o-isomorphic to a subdirect sum of the f. o. rings $R/P_i = R_i$. These again satisfy the minimum condition, therefore they contain minimal L-ideals. From Theorem 8 of Chapter VIII we infer that the R_i are o-simple.

It remains to verify that a direct sum may be obtained from the subdirect sum. Supposing that the P_i are all different and by making use of the distributivity of the lattice of L-ideals, we see that

$$(P_1 \cap \ldots \cap P_{i-1}) + P_i = (P_1 + P_i) \cap \ldots \cap (P_{i-1} + P_i) = R.$$

This implies at once that R is the direct sum of the R_i.[5]

There is clearly no fundamental difficulty in carrying over the other kinds of radicals, used in ring theory with considerable success, to the case of l. o. rings. One has merely to add the postulate that all ideals should be L-ideals. But these radicals have not led to structural theorems on l. o. rings in general, only in the case of F-rings (mentioned in **2**), so that we shall not enter into their discussions.

For further results on l. o. rings we refer to BIRKHOFF—PIERCE [1] and D. G. JOHNSON [1].

---

[5] Note that the P_i are maximal L-ideals in R.

PARTIALLY ORDERED SEMIGROUPS

CHAPTER X

PARTIAL ORDERS ON SEMIGROUPS

1. Partially ordered groupoids and semigroups

By a *partially ordered groupoid* (*p. o. groupoid*) we understand a set H satisfying

S 1. H is a groupoid, i. e. H is closed under a multiplication;

S 2. H is a p. o. set under a relation $\leq$;

S 3. $a \leq b$ implies $ac \leq bc$ and $ca \leq cb$ for all $c \in H$.

If we assume that the multiplication is associative, we obtain a *p. o. semigroup*. We preserve the terminology and notation for the different kinds of partial orders, so that (e. g.) the meaning of f. o. groupoid or directed semigroup will be obvious. The reader should perhaps be reminded at this point that the existence of inverses excludes several special phenomena in p. o. groups which may well occur in p. o. semigroups; thus e. g. a p. o. semigroup may be a $\vee$-semilattice without being a lattice; and $a < b$ is consistent with $ac = bc$, etc. Let us therefore begin with some new terminology.

A p. o. semigroup which is a lattice under its partial order, does not necessarily satisfy the useful distributive laws $c(a \vee b) = ca \vee cb$ etc. Little generality is lost if we define, with a view to applications, a *l. o. groupoid* as a groupoid H which is a lattice satisfying

$$(a \vee b)c = ac \vee bc, \quad c(a \vee b) = ca \vee cb$$

for all $a, b, c \in H$. If H is a p. o. groupoid and at the same time a $\vee$- or $\wedge$-semilattice under its partial order, we write briefly: H is a $\vee$-*groupoid* ($\wedge$-*groupoid*).

If H obeys a stricter set of axioms than those above, namely, if S3 is replaced by

S 3*. $a < b$ implies $ac < bc$ and $ca < cb$ for all $c \in H$,

then we say that the partial order of H is *strict*, or H is a *strict p. o. groupoid*. It is an easy matter to prove that a p. o. groupoid

H is·strict if, and only if, it satisfies a mild cancellation law: $ac = bc$ (or $ca = cb$) implies $a = b$ or $a \parallel b$. If H is a f. o. groupoid, S3* is equivalent to S3 plus the cancellation laws in H.

A p. o. groupoid H will be called *strong* if

$$ac \leq bc \quad (\text{or } ca \leq cb) \quad \text{implies} \quad a \leq b.$$

A strong p. o. groupoid obviously always obeys the cancellation laws and is therefore strict. In f. o. groupoids the concepts "strict" and "strong" are equivalent.

Call an element a of a p. o. groupoid H *l-positive, r-positive* or *positive* according as $ax \geq x$, $xa \geq x$ or both hold for all $x \in H$. *l-negative, r-negative* and *negative* elements are defined by the dual inequalities. *Strictly l-positive* etc. elements may be similarly defined by strict inequalities. The corresponding sets are denoted by P_l, P_r, P, N_l, N_r, N, and by P_l^*, P_r^*, P^*, N_l^*, N_r^*, N^* in the strict case. They will be discussed in **3** for the case of p. o. semigroups.

A p. o. groupoid will be called *positively* (*negatively*) *ordered* if all of its elements are positive (negative).

We say that a p. o. groupoid H is *naturally ordered* if it is positively ordered and

$$a < b \quad \text{implies} \quad ax = ya = b \quad \text{for some} \quad x, y \in H.$$

Proposition 1. (NAKADA [1].) *A necessary and sufficient condition that a p. o. semigroup S be the positive cone of a p. o. group G is that it satisfy*

(a) *S is cancellative (i. e. it satisfies both cancellation laws),*

(b) *S contains a neutral element,*

(c) *S is naturally ordered.*

Necessity is obvious. For sufficiency we verify the conditions (iii), (iv) of Theorem 4 in Chapter II. Because of (c), $a, b \in S$ satisfy $ab \geq a = ae \geq e$. Therefore, $ab = e$ implies $ab = a = e$ and (iii) holds. Since for $b \in S$, $b > e$, we have $ab > a$, again (c) implies $ab = ca$ for some $c \in S$. Therefore, $aS \subseteq Sa$, and by symmetry we obtain (iv). It is obvious that the partial

order of S coincides with that induced by the partial order of
the group $\{S\}$ with S as positive cone.

The special case when S is l. o. has been considered by
DUBREIL [2]. By his result conditions (a)—(c) can be combined
into the single one: $a, b \in S$ and $a \leq b$ imply the existence of
a unique $c \in S$ and a unique $d \in S$ such that $b = ac = da$.
Of course, it must also be assumed that S is positively ordered.

Since the multiplicative groupoid of a p. o. ring is not a p. o.
groupoid in the above sense (S3 is satisfied only by positive elements),
it seems natural to investigate groupoids and semigroups satisfying
only S1 and S2. This has been begun by TAMARI [1] and continued
by CLIFFORD [6]. CLIFFORD considers semigroups which are f. o. as
sets and in which every element either preserves or reverses partial
order when inequalities are multiplied by it.

2. Examples

In this section we shall give some illustrative examples of
p. o. groupoids and p. o. semigroups.

1. Let H be a (closed or open) interval of the real line with
the usual ordering relation, and consider H as a groupoid
under the operation of forming weighted arithmetic means:

$$a \cdot b = \lambda a + \mu b$$

where λ, μ are fixed real numbers satisfying $\lambda > 0$, $\mu > 0$,
$\lambda + \mu = 1$. Then H is a strong f. o. groupoid.

2. Again let H mean an interval of the real line and f
a function from $H \times H$ into H which is not decreasing in any
variable. Then H endowed with f as a binary operation is
a f. o. groupoid. H is strong exactly if f is strictly increasing
in both variables.

3.[1] Consider the set H of all normal subgroups of some
group G. If the "product" of $A, B \in H$ is defined as the commutator $[A, B]$ of A and B, and if $\leq$ means set-inclusion,
H becomes a l. o. groupoid.

4. The multiplicative semigroup of all positive integers will
be a strong l. o. semigroup if $a \leq b$ means $a \mid b$ (divisibility).
(This is naturally ordered.)

[1] Examples 3, 7 and 8 are due to BIRKHOFF [3].

5. In a p. o. ring R the positive cone R^+ is a commutative p. o. semigroup under addition and a p. o. groupoid under multiplication. The latter is strict if, and only if, R^+ contains no divisors of zero.

6. Let S be the multiplicative semigroup of all ideals of an associative ring R and let $\leq$ mean inclusion. Then S is a l. o. semigroup.

7. Let S be the semigroup of all o-preserving mappings of a p. o. set T into itself, and define $\alpha \leq \beta$ ($\alpha, \beta \in S$) to mean that $\alpha(a) \leq \beta(a)$ for all $a \in T$. Then S will be a p. o. semigroup.

8. The $\vee$-endomorphisms of a $\vee$-semilattice L form a $\vee$-semigroup if multiplication is interpreted as the usual endomorphism multiplication and the join $\theta_1 \vee \theta_2$ of two endomorphisms is defined by

$$(\theta_1 \vee \theta_2)(x) = \theta_1(x) \vee \theta_2(x).$$

It is routine to check the requisite postulates.

9.[2] The subsets $A, B, \ldots$ of a semigroup S under complex multiplication and set-inclusion form a l. o. semigroup. Here the infinite distributive laws

$$A(\cup B_\nu) = \cup(AB_\nu) \quad \text{and} \quad (\cup B_\nu)A = \cup(B_\nu A)$$

hold, but the analogous laws for $\cap$ do not hold in general (only the inclusions $A(\cap B_\nu) \leq \cap(AB_\nu)$, $(\cap B_\nu)A \leq \cap(B_\nu A)$ do).

10. Every lattice can be made into a l. o. semigroup by introducing the trivial (l. u. b.) multiplication:

$$ab = ba = a \vee b.$$

11.[3] We mention the following real intervals I under the usual ordering and with the indicated operations. They are all f. o. semigroups:

---

[2] This important example has been discussed in detail by DU-BREIL [2].

[3] Examples (a) and (b) play an important part in the theory of Archimedean f. o. semigroups (CLIFFORD [5]; cf. Theorem 2 in Chapter XI). Examples (c) are due to HION [1]; the second one is a non-nilpotent semigroup all of whose elements are nilpotent.

(a) $I_1 = (0, 1]$ where $a \cdot b = \min(a + b, 1)$,

(b) $I_2 = (0, 1] \cup \infty$ where $a \cdot b = a + b$ if $a + b \leq 1$ and $a \cdot b = \infty$ if $a + b > 1$,

(c) $I_3 = [\frac{1}{2}, 1]$ or $[\frac{1}{2}, 1)$ where $a \cdot b = \max(\frac{1}{2}, ab)$.

12.[4] Let S be the set of all monotone functions from $[0,1]$ into $[0, 1]$ where the product $f \cdot g$ of f and g is defined as $(f \cdot g)(x) = f(g(x))$ for all x, and $f \leq g$ means $f(x) \leq g(x)$ for all x. This is a l. o. semigroup.

13. The real numbers in $[0, 1]$ under the operation

$$a \cdot b = a + b - ab \quad \text{or} \quad a \times b = \frac{a+b}{1+ab}$$

form a f. o. semigroup.

3. The positive and negative cones

In this section we are concerned with the sets P_l, P_r, P, $N_l, \ldots$ and P_l^*, P_r^*, $P^*, \ldots$ in p. o. semigroups S as defined in 1. They will be called the *l-positive cone, ... l-negative cone, ...* of S. Here we list their most important properties, some of which are so obvious that the proofs may be omitted.

A) The following relations hold:

(a) $P_l^* \subseteq P_l$, $P_r^* \subseteq P_r$, $P^* \subseteq P$, $N_l^* \subseteq N_l$, $N_r^* \subseteq N_r$, $N^* \subseteq N$,

(b) $P = P_l \cap P_r$, $P^* = P_l^* \cap P_r^*$, $N = N_l \cap N_r$, $N^* = N_l^* \cap N_r^*$,

(c) $P_l \cap N_l^* = P_l^* \cap N_l = P_r \cap N_r^* = P_r^* \cap N_r = P \cap N^* = P^* \cap N = \varnothing$.

B) *All the cones are subsemigroups of S.*

C) Any one of the positive cones contains together with an element a all elements x of S satisfying $x \geq a$. The dual statement holds for the negative cones. Thus *all the cones are convex.*

D) P_l^* $[N_l^*]$ *is a right ideal of P_l $[N_l]$ and P_r^* $[N_r^*]$ is a left ideal of P_r $[N_r]$*; for if $a \in P_l^*$ and $b \in P_l$, then $bx \geq x$ and so $abx \geq ax > x$ for all $x \in S$.

E) We have

$$P_l P_r \subseteq P, \quad P_l^* P_r^* \subseteq P^*, \quad N_l N_r \subseteq N, \quad N_l^* N_r^* \subseteq N^*.$$

[4] See BIRKHOFF [3] and DUBREIL-JACOTIN—LESIEUR—CROISOT [1].

In fact, if $a \in P_l$, $b \in P_r$, then $ab \geq a$ implies $abx \geq ax \geq x$ and $ab \geq b$ implies $xab \geq xb \geq x$ for all $x \in S$. — Hence we also conclude:

F) P *is a left ideal of* P_l *and a right ideal of* P_r. Similar assertions hold for P^*, P_l^*, P_r^* and for the negative cones.

G) $P_l \cap N_l$ is the set of all left identities, $P_r \cap N_r$ the set of all right identities of S and $P \cap N$ is the identity e of S (if it exists).

H) For all $a \in P_l$, $b \in N_r$ and for all $a \in P_r$, $b \in N_l$ the inequality $a \geq b$ holds, because $a \geq ab \geq b$ or $a \geq ba \geq b$. Hence $P_l \cap N_r$ contains at most one element c and $P_r \cap N_l$ contains at most one element d (which are idempotent and satisfy $xcy = xy = xdy$ for all $x, y \in S$).

I) A notable simplification arises if we assume the existence of an identity in S.

Proposition 2. *If* S *contains an identity* e, *then* $P_l = = P_r = P$ *and* $N_l = N_r = N$. *Further,* $a \in S$ *belongs to* P *(to* N) *if, and only if,* $a \geq e$ $(a \leq e)$.

In fact, if $a \in P_l$, then because of H) and $e \in N_r$ we get $a \geq e$. That $a \geq e$ implies $a \in P$ is obvious.

J) It is a well-known result that every semigroup S can be embedded in a semigroup with identity. For p. o. semigroups the corresponding result reads as follows.

Theorem 3. *A p. o. semigroup* S *without identity element can be embedded in a p. o. semigroup* $\bar{S}$ *with identity* e *with preservation of ordering, l- and r-positivity, l- and r-negativity, if, and only if, the cones of* S *satisfy* $P_l = P_r$ *and* $N_l = N_r$. *The minimal* $\bar{S}$'s *are o-isomorphic over* S.

If S is embeddable in the desired way, say in $\bar{S}$, then it is so embeddable in $S \cup e$, so that we may assume $\bar{S} = S \cup e$. Since the cones of $\bar{S}$ are the cones of S with e adjoined, the necessity follows at once from Proposition 2. Conversely, let the stated condition be satisfied. In $S \cup e = \bar{S}$ we define as new ordering relations:[5] $a > e$ for all $a \in P$ and $b < e$ for all $b \in N$ (which hold necessarily in $\bar{S}$ in view of H)). These new

---

[5] P, N denote the positive and negative cones of S, respectively.

relations do not imply, neither by transitivity, nor by multiplication by elements of S, any new ordering relation between elements of S, thus $\bar{S}$ is of the stated kind. The o-isomorphy of any two $S \cup e$ follows from the fact that $a > e$ ($a < e$) in $S \cup e$ if, and only if, $a \in P$ ($a \in N$).

K) Let $a \in P_l \cup P_r$. Then either $a < a^2 < \ldots < a^n < < a^{n+1} < \ldots$ or $a < a^2 < \ldots < a^n = a^{n+1} = \ldots$. If $a \in P_l^* \cup P_r^*$, then only the first alternative can occur. This holds as $aa^k \geq a^k$ or $a^k a \geq a^k$ for every k, and the inequality is strict if $a \in P_l^* \cup P_r^*$. — If $a \in N_l \cup N_r$, then $a > a^2 > \ldots > a^n > > a^{n+1} > \ldots$ or $a > a^2 > \ldots > a^n = a^{n+1} = \ldots$.

L) The set P_l has the property: if $a, b \in P_l$ and ab is a left identity, then both a and b are left identities. $bx \geq x$ implies $x = abx \geq ax$ whence $ax = x$ for all $x \in S$. Therefore $a \in P_l \cap \cap N_l$ and so $b = ab \in P_l \cap N_l$. Similar properties hold for P_r, P and N_l, N_r, N.

M) If S is cancellative, then

$$P_l \setminus P_l^* = P_r \setminus P_r^* = P \setminus P^* = N_l \setminus N_l^* = N_r \setminus N_r^* = N \setminus N^*;$$

this set consists of the identity e of S if it exists, and it is empty otherwise. In fact, let e. g. $a \in P_l \setminus P_l^*$; then $ax_0 = x_0$ for some $x_0 \in S$, and so $yax_0 = yx_0$, $ya = y$ for all $y \in S$. Thus a is a right identity and analogously a left identity of S, so $a = e$.

N) Let now S be strong. $a \in P_l^*$ (or $a \in P_r^*$) if, and only if, $a < a^2$; $b \in N_l^*$ (or $b \in N_r^*$) if, and only if, $b^2 < b$. By K), only sufficiency need be verified. But this follows at once since $a < a^2$ implies $ax < a^2 x$ whence $x < ax$ for all $x \in S$, $a \in P_l^*$. In particular, we obtain $P_l = P_r = P$ and $N_l = N_r = N$. — A similar reasoning applies to prove:

O) In a strong p. o. semigroup S, $a \in P_l^*$ (or $a \in P_r^*$) if, and only if, there exists an $x_0 \in S$ such that $ax_0 > x_0$ or $x_0 a > x_0$, and dually for N_l^* and N_r^*.

P) In a strict f. o. semigroup S, the relation $S = P \cup N$ holds. For $ax_0 > x_0$ and $x_1 a \leq x_1$ (for some $x_0, x_1 \in S$) lead to the contradictory inequalities $x_1 ax_0 > x_1 x_0$, $x_1 ax_0 \leq \leq x_1 x_0$.

4. Semigroups of quotients

The fundamental difference between the theory of p. o. semigroups and those of p. o. groups and rings stems from the fact that in the latter the sets of positive elements characterize the partial orders, whereas this is not the case in semigroups. In fact, the results of Chapters III and VII have no analogues for semigroups.

There exists, however, an obvious analogue of the method of ordering a ring of quotients. It shows how to extend the full order of a semigroup S to its group G of quotients when G exists. The results obtained there extend readily to our present case, and therefore we shall begin with a more general case.

Let S be a semigroup and T a semigroup containing S. Assume that to every $a \in T \setminus S$ there exist elements a, b in S such that a is left-cancellable,[6] b is right-cancellable in T and aa, ab are in S. Then we say that T is *a semigroup of quotients of S*.

Theorem 4. (FUCHS [11].) *A full order of a semigroup S can be extended, in one, and only one, way, to a full order of a semigroup T of quotients of S.*

If T properly contains S, then there exist elements $a, b \in S$ which are left- and right-cancellable, respectively, in T, and therefore, in the above definition, the case $a \in S$ need not be excluded. Now if $a, \beta \in T$ ($a \neq \beta$) and if a, b are left- and right-cancellable elements such that $aa, \beta b \in S$, then aab and $a\beta b$ are different elements of S and we define $a \gtrless \beta$ according as $aab \gtrless a\beta b$ in S. For the elements of S this definition evidently coincides with that originally given in S. The definition does not depend on the choice of a, b; for if a', b' are again left- and right-cancellable elements with $a'a$, $\beta b' \in S$, then—taking a left- and right-cancellable element $a'', b'' \in S$ such that $a''(a'\beta), (ab)b'' \in S$—we obtain (e. g.) from $aab < a\beta b$ in turn $aabb'' < a\beta bb''$, $abb'' < \beta bb''$, $a''a'abb'' <$ $< a''a'\beta bb''$, $a''a'a < a''a'\beta$, $a''a'ab' < a''a'\beta b'$, $a'ab' < a'\beta b'$. A similar argument applies if a', b' are determined so as to

---

[6] a is called left-cancellable if $ax = ay$ implies $x = y$.

have $a'\beta$, $ab' \in S$. The transitivity of $<$ follows by a straight-forward computation of a similar kind. Finally we show that $a \leq \beta$ implies $\gamma a \leq \gamma \beta$ for all $\gamma \in T$. If a, b, b'' are defined as before and $c \in S$ is left-cancellable such that $c\gamma \in S$, then we get successively $aab \leq a\beta b$, $aabb'' \leq a\beta bb''$, $abb'' \leq \beta bb''$, $c\gamma abb'' \leq c\gamma \beta bb''$. Hence by multiplying by suitable elements, we arrive at $\gamma a \leq \gamma \beta$. The uniqueness is evident.

By specializations we obtain the following corollaries :

Corollary 5. (CONRAD [11].) *Let S be a cancellative f. o. semigroup, and let every pair of elements a, b of S have a common right multiple $ax = by$. Then S can be embedded in a f. o. group G of quotients $g = ab^{-1}$ (a, $b \in S$) such that*

$$g > e \text{ if, and only if}, a > b \text{ in } S.$$

This G is unique to within o-isomorphism.

The purely algebraic part of this result is well known, so that the existence of the group G of quotients of S may be taken for granted. The remaining part is an immediate consequence of Theorem 4.

It is easy to see that S is contained in the positive cone P of G if, and only if, S is positively ordered; and $S = P$ or $S = P \setminus e$ if, and only if, S is naturally ordered (cf. Proposition 1).

Note that if a cancellative semigroup S is either naturally ordered, or merely if $a < b$ in S implies that $b = ax$ for some $x \in S$, then it satisfies the hypotheses of Corollary 5, and so it admits a f. o. group of quotients. In fact, $a(xy) = by$ for all $y \in S$.

Corollary 6. (TAMARI [1], ALIMOV [1], NAKADA [1].) *A cancellative commutative f. o. semigroup S is embeddable in a f. o. Abelian group G, unique up to o-isomorphism, such that every element of G is the quotient of two elements of S.*

A classical result of A. I. MAL'CEV states that not every cancellative semigroup can be embedded in a group. CHEHATA [2] and VINOGRADOV [3] considered, simultaneously and independently, the corresponding question for f. o. semigroups. They proved that the example given by MAL'CEV can be f. o.; thus there exist f. o. semigroups that cannot be embedded in groups.

CHAPTER XI

FULLY ORDERED SEMIGROUPS

1. Definitions and preliminary lemmas

We begin the theory of f. o. semigroups with a systematic treatment of the Archimedean case. As one naturally expects, deeper results can be achieved only if some additional postulates are imposed on the semigroup, like natural order or cancellability. The main results are to be found in **2** and **3**.

We start with two fundamental definitions. A f. o. semigroup S will be called *Archimedean* if

1) $a, b \in P$ and $a^n < b$ for all positive integers n imply $a = e$ (= the identity), and

2) $a, b \in N$ and $a^n > b$ for all $n > 0$ imply $a = e$.

Here and in the sequel, P (P^*) and N (N^*) denote the (strict) positive and negative cones of S, respectively.

Two distinct elements a, b of S are said to form an *anomalous pair*[1] if

$$a^n < b^{n+1}, \quad b^n < a^{n+1} \quad \text{for all } n > 0,$$

or

$$a^n > b^{n+1}, \quad b^n > a^{n+1} \quad \text{for all } n > 0.$$

Clearly, the first alternative may occur if $a, b \in P$ and the second if $a, b \in N$.

The following three lemmas are mainly of technical character.

Lemma A. *Let S be a f. o. semigroup, $a, b \in P$ and $ab \leq \leq ba$. Then*

(1) $$a^n b^n \leq (ab)^n \leq (ba)^n \leq b^n a^n,$$

(2) $$a^n b^m \leq b^m a^n$$

for all natural integers n, m. If, in addition, S is Archimedean, then for every n

(3) $$(ba)^n \leq a^{n+1} b^n \leq b^n a^{n+1} \leq (ab)^{n+1}.$$

---

[1] For this concept see ALIMOV [1].

(1) trivially holds for $n = 1$. We proceed by induction and suppose that (1) holds for $n - 1$. Then $a^n b^n = a(a^{n-1} b^{n-1})b \leq$ $\leq a(ba)^{n-1} b = (ab)^n$, and similarly $(ba)^n \leq b^n a^n$. Inequality (2) follows by a similar induction. To prove (3), suppose S Archimedean and $a \neq e$. Define the integer $k = k(n)$ so as to satisfy $a^{k-1} \leq b^n \leq a^k$. We obtain $(ba)^n \leq b^n a^n \leq a^{k+n} \leq$ $\leq a^{n+1} b^n$, and analogously for the last inequality in (3).

The next lemma describes the relationship between the two notions defined above.

Lemma B.[2] *Let S be a cancellative f. o. semigroup. If S has no anomalous pair, then it is Archimedean. If S is Archimedean and naturally ordered, then it contains no anomalous pair.*

Let $a, b \in P$ satisfy $a^n < b$ for all $n > 0$; and let $a \neq e$. If $ab \leq ba$, then $b^n < (ba)^n < a(ba)^n b = (ab)^{n+1}$, and, on account of (1), $(ab)^n \leq b^n a^n < b^{n+1}$. Thus b and ab form an anomalous pair. If $ba \leq ab$, then b and ba yield such a pair. The case $a, b \in N$ is symmetrical.

Assume that S is Archimedean and naturally ordered. If $a < b$ in S, then there exist $c, d \in S$ such that $b = ca = ad$. Let e. g. $c \geq d$. Then $ac \geq ca$, and (1) implies $b^n = (ca)^n \geq$ $\geq c^n a^n$. The integer n may be chosen such that $c^n > a$, and then we arrive at $b^n > a^{n+1}$. Consequently, the pair a, b cannot be anomalous.

It is easy to show that an Archimedean cancellative f. o. semigroup may contain anomalous pairs even if commutativity is assumed.[3] Let the semigroup S consist of all $a^n b^m$ with $n, m \geq 0$ and with multiplication

$$a^n b^m \cdot a^k b^l = a^{n+k} b^{m+l}.$$

Define $a^n b^m < a^k b^l$ if either $n + m < k + l$ or $n + m = k + l$ and $n > k$. The pair a, b is anomalous.

Lemma C. (CLIFFORD [5].) *Let S be an Archimedean, naturally f. o. semigroup. If S is not cancellative, then*

(a) *it contains a maximal element u,*

(b) *for every $a \in S$ $(a \neq e)$, there exists a natural integer k such that $a^k = u$,*

(c) *$ab = ac \neq u$ (or $ba = ca \neq u$) implies $b = c$.*

---

[2] The first part of this lemma is due to ALIMOV [1].
[3] See CLIFFORD [7].

By hypothesis, three elements $a, b, c \in S$ exist such that $ab = ac$ and $b < c$ (or $ba = ca$ and $b < c$). Since $c = bx$ for some $x \in S$ ($x \neq e$), $u = ab$ satisfies $u = ux$. If there existed an element $y \in S$ greater than u, then choosing n so as to satisfy $x^n \geq y$, we should have $u = ux^n \geq uy \geq y > u$, a contradiction. This establishes (a) and (c) at once. By the Archimedean property, $a \neq e$ implies $a^k \geq u$ for some k, and so $a^k = u$. Q. E. D.

We note in passing that if $a < b \neq u$ in the semigroup S of Lemma C, then there exists exactly one $c \in S$ satisfying $b = ac$.

2. Archimedean, naturally fully ordered semigroups

If the Archimedean f. o. semigroups are supposed to be naturally ordered, then they can be described with some precision. This might be expected in the cancellative case (when the f. o. semigroups are actually the "halves" of f. o. groups, at least if groups of quotients exist), but surprisingly, it is so even in the non-cancellative case.

First we establish the important commutativity theorem:

Theorem 1. (Hölder [1], Fuchs [10].) *An Archimedean, naturally f. o. semigroup is commutative.*

For the sake of convenience, and without loss of generality, we omit the identity from the semigroup if it has one.

First, assume that the semigroup S possesses a minimal element a. Let $b \in S$ and assume $b \neq a$. Then there exists an integer $k \geq 1$ such that $a^k < b \leq a^{k+1}$, and an element $c \in S$, $c \geq a$, such that $b = a^k c$. Now $a^{k+1} = a^k a \leq a^k c = b \leq \leq a^{k+1}$. Hence b is a power of a, and S consists of the positive powers of a.

Next assume that S has no minimal element.[4] Then to every $x \in S$ there exists a $z \in S$ with $z^2 \leq x$; indeed, if $y < x$ and $x = yy_0$, then $z = \min(y, y_0)$ is such an element. By way

---

[4] This part of the proof is similar to that of Theorem 1 in Chapter IV. However, certain complications arise, because we cannot use an argument based on "large" elements; instead we argue with "small" elements which are at our disposal.

of contradiction, suppose that $ba < ab$ for some $a, b \in S$. First let $ab < u$. If $ab = bax$ and $z^2 \leq x$, $z \leq a$, $z \leq b$, then we determine integers m, n satisfying $z^m \leq a < z^{m+1}$, $z^n \leq b < z^{n+1}$. But these lead to the inequalities $ab = bax \geq z^{m+n+2} > ab$ (in the non-cancellative case strict inequality because of Lemma C), a contradiction. Next let $ab = u$, and say $a < b$. Then $a^k < b \leq a^{k+1}$ for some $k \geq 1$. Therefore $b = a^k c$ for a certain $c \leq b$, and since $ac \leq b$ and $ca \leq ba < ab = u$, we can apply what has already been proved to conclude that a and c commute. This is again a contradiction, for then a and b also commute. Consequently, S is commutative.

The following result characterizes the Archimedean f. o. semigroups in case they are naturally ordered.

Theorem 2. (HÖLDER [1], CLIFFORD [5].) *Let S be an Archimedean, naturally f. o. semigroup. Then S is o-isomorphic to a subsemigroup of one of the following f. o. semigroups:*

P: *the additive semigroup of all non-negative real numbers;*

P_1: *the real numbers in the interval $[0, 1]$ with the usual ordering and $ab = \min(a + b, 1)$;*

P_2: *the real numbers in $[0, 1]$ and the symbol ∞ with the usual ordering and the operation $ab = a + b$ if $a + b \leq 1$ and $ab = \infty$ if $a + b > 1$.*

The first case occurs if, and only if, S is cancellative.

From the preceding theorem we know that S is necessarily commutative.

Let S be generated by a single element a. Then its elements are $(e <) a < a^2 < \ldots < a^n < \ldots$ or $(e <) a < \ldots < a^n = a^{n+1}$. The mapping

$$a^k \to k \quad \text{or} \quad a^k \to \frac{k}{n}$$

embeds S o-isomorphically in P or in P_1.

If[5] S is not generated by a single element, then we again discard the identity of S if it exists. To every $x \in S$ there exists a $z \in S$ such that $z^2 \leq x$ (cf. the proof of the preceding

---

[5] For the next proof see FUCHS [10]. Cf. footnote[4].

theorem); hence also, if a positive integer t is given, there exists a $z \in S$ such that $z^t \leq x$. We now choose an element $a \in S$ less than the maximal element u of S if S has a maximal element, but otherwise arbitrary, and we define a function f on S as follows.[6] To every $b \in S$, $b < u$, we form two classes of pairs of positive integers:

$$L_b = [(m, n) \mid a \leq x^n \text{ and } x^m \leq b \text{ for some } x \in S]$$

and

$$U_b = [(k, l) \mid b \leq y^k \text{ and } y^l \leq a \text{ for some } y \in S].$$

The Archimedean character of S guarantees that neither L_b nor U_b is empty. Next we show that

(*) $$\frac{m}{n} \leq \frac{k}{l} \quad \text{for all } (m, n) \in L_b \text{ and } (k, l) \in U_b.$$

To every large t we can find a $z \in S$ with $z^t \leq \min(x, y)$. Hence $r \geq t$ and $s \geq t$ hold for the integers r, s defined by $z^r \leq x < z^{r+1}$, $z^s \leq y < z^{s+1}$. Therefore

$$z^{rm} \leq b < z^{(s+1)k} \quad \text{and} \quad z^{sl} \leq a < z^{(r+1)n}$$

whence $rm < (s + 1)k$ and $sl < (r + 1)n$. We infer that

$$\frac{m}{n} < \left(1 + \frac{1}{r}\right)\left(1 + \frac{1}{s}\right)\frac{k}{l}$$

for arbitrarily large r, s, thus (*) holds. On the other hand, a similar argument shows that to every large $t > 0$ there exist $r, s \geq t$ such that $(s, r + 1) \in L_b$ and $(s + 1, r) \in U_b$. Since the difference is

$$\frac{s + 1}{r} - \frac{s}{r + 1} \to 0$$

with increasing t, it follows that there exists one, and only one, real number β such that

$$\frac{m}{n} \leq \beta \leq \frac{k}{l} \quad \text{for all } (m, n) \in L_b \text{ and } (k, l) \in U_b.$$

We put $f(b) = \beta$; then $f(a) = 1$.

[6] f is the symbolic logarithm to base a.

There is no difficulty in proving that the function f from $S\setminus u$ to the real axis is isotone and satisfies $f(b) > 0$ for all $b \in S\setminus u$. Moreover we have $f(bc) = f(b) + f(c)$ whenever $bc < $ $< u$ — which can again be proved by using sufficiently small elements $z \in S$. Hence $f(b) = f(c)$ only if $b = c$.

If S is cancellative, then it contains no maximal element and f is an o-isomorphism of S into P. If S is not cancellative, then in view of Lemma C it contains a maximal element u. The set of values of $f(b)$ (for all $b \in S\setminus u$) is bounded: if $a^k = u$, then k is an upper bound. Thus there exists a smallest real number a such that $f(b) \leq a$ for all $b \in S\setminus u$. If no $c \in S\setminus u$ exists with $f(c) = a$, then set $f(u) = a$, and if such a c exists, then $f(u) = \infty$. The function

$$g(b) = \frac{1}{a} f(b)$$

is obviously an o-isomorphism of S into P_1 or P_2. This completes the proof.

It is easy to conclude that any two o-isomorphisms of S into P differ merely in a positive real factor, while those into P_1 or P_2 are necessarily identical.

Corollary 3. (HUNTINGTON [1], [2].) *An Archimedean naturally f. o. semigroup S which is cancellative and has a least element but no identity is o-isomorphic to the additive semigroup of the natural numbers.*

The least element a of S satisfies $a < a^2 < \ldots < a^n < \ldots$. If $a^n < x \leq a^{n+1}$ for some $x \in S$, then $x = ya^n$ for $y \in S$, $y \leq a$ whence $x = a^{n+1}$ and S consists of the powers of a.

3. Subsemigroups of the group of real numbers

The question to which we must now turn is to characterize by simple conditions the f. o. semigroups that can be embedded o-isomorphically in the additive group of all real numbers. The next theorem shows in what circumstances this is possible.

Theorem 4. (ALIMOV [1].) *A f. o. semigroup S is o-isomorphic to a subsemigroup of the additive group of the real numbers if, and only if, it satisfies the conditions:*

(a) S *contains no anomalous pairs,*

(b) S *is cancellative.*

It is obvious that the subsemigroups of the real group satisfy both (a) and (b), so that we may turn immediately to the proof of sufficiency. Assuming S is a f. o. semigroup satisfying (a) and (b), we conclude in turn:

1°. S is Archimedean because of Lemma B in **1**.

2°. S is commutative. First let $a, b \in P^*$ and, by way of contradiction, $ab < ba$. Then

$$(ab)^n < (ba)^{n+1}, \quad (ba)^n < a(ba)^n b = (ab)^{n+1}.$$

Hence ab, ba form an anomalous pair, a contradiction to (a). Thus $ab = ba$. If $a, b \in N^*$, the proof of $ab = ba$ is similar. Finally, let $a \in P^*$, $b \in N^*$. If $ab \in P$, then because of what has already been proved, $a(ab) = (ab)a$ whence (b) implies $ab = ba$. If $ab \in N$, then $b(ab) = (ab)b$ implies $ba = ab$.

3°. The group G of quotients of S is Archimedean. From Corollary 6 of Chapter X we know that G can be f. o. We first show that if P^* is not void, then the elements of G may be written in the form ab^{-1} with $a, b \in P$. This will follow at once if we can show that to every $x \in N^*$ there exists a $y \in P$ such that $xy \in P$. If to $x \in N^*$ no such $y \in P$ existed, then $xy^n \in N^*$ would hold for $n = 1, 2, \ldots$ and for all $y \in P^*$. Since

$$(xy)^n > x^n > x^{n+1}, \quad x^n > x^n(xy^{n+1}) = (xy)^{n+1},$$

the pair xy, x would be anomalous, contrary to (a). Now, if G is not Archimedean, then we can find elements $a, b, c, d \in S$, all in P or all in N, such that $c < a$, $d < b$ and $(ac^{-1})^n < bd^{-1}$ for every n. If all lie in P, then the pair bc, ba is anomalous, for

$$(bc)^n < (ba)^{n+1}, \quad (ba)^n = b^n a^n < b^{n+1} d^{-1} c^n \leq (bc)^{n+1},$$

while if all belong to N, then ad and cd form an anomalous pair:

$$(ad)^n > (cd)^{n+1}, \quad (cd)^n > a^n b^{-1} d^{n+1} \geq (ad)^{n+1}.$$

Thus G is Archimedean, as stated.

It is now easy to complete the proof of the theorem. S has a group of quotients which is an Archimedean f. o. group. By Hölder's Theorem 1 of Chapter IV, the proof is finished.

If we restrict ourselves to positively (or negatively) ordered semigroups, then the rather heavy condition of being cancellative may be replaced by two weaker ones:

Theorem 5. (Fuchs [10].) *A necessary and sufficient condition that a positively f. o. semigroup S be o-isomorphic to a subsemigroup of the real group is that it satisfy*

(a) *S contains no anomalous pair,*

(b) *S is Archimedean,*

(c) *S contains no maximal element unless it consists of a single element.*

Here again only the sufficiency needs a verification. Supposing (a), (b) and (c), we get successively:

$1°$. If $a, b \in S$ and $a \neq e$, then $ab > b$. For let $c > b$ and choose n so large that $a^n \geq c$. Then $ab = b$ would imply $a^n b = = b$ whence $a^n b \geq a^n \geq c > b$, a contradiction. Thus $a \neq e$ implies $a \in P^*$.

$2°$. S is commutative. If $ab \neq ba$ for $a, b \in P^*$, then $(ba)^n < < a(ba)^n b = (ab)^{n+1}$ and $(ab)^n < (ba)^{n+1}$ for all $n > 0$, contrary to (a).

$3°$. S is cancellative. Assume that $ab = ac$ and $b < c$. By making use of commutativity, a simple induction shows that $ab^n = ac^n$ for all $n > 0$. By $1°$ then $ac^n < ac^{n+1} = ab^{n+1}$, and hence $c^n < b^{n+1}$ for all n. Obviously $b^n \leq c^n < c^{n+1}$. Condition (a) shows that the assumption is contradictory, and thus S is cancellative.

The required conclusion is now an immediate consequence of Theorem 4.

4. Archimedean semigroups with anomalous pairs

So far we have considered Archimedean f. o. semigroups subject either to the condition of absence of anomalous pairs or to that of being naturally ordered. Without either of these hypotheses we are able to prove only rather weak statements.

Let S be an Archimedean f. o. semigroup and suppose that it is positively ordered.

If $a < b < c$ and a, c form an anomalous pair, then so do the pairs a, b and b, c. The converse is also true: if a, b and b, c are anomalous pairs, then so is a, c. For if $a^{n+1} \leq c^n$ for some n, then also $a^{2n+2} \leq c^{2n} < b^{2n+1}$ in contradiction to $b^{2n+1} < a^{2n+2}$. We infer that S splits into disjoint intervals I_λ (some of them may collapse to a single element) such that two distinct elements are anomalous exactly when they belong to the same interval.

If a, b form an anomalous pair, then, for all $c \in S$, so do the pairs ac, bc and ca, cb provided S is cancellative. In fact, $a < b$ implies $ac < bc$ whence $(ac)^n < (bc)^{n+1}$ for all n, and from (1) and (3) of Lemma A in 1 we obtain:

1. if $ac \leq ca$ and $bc \leq cb$, then
$$(bc)^n \leq c^n b^n < c^n a^{n+1} \leq (ac)^{n+1};$$

2. if $ac \leq ca$ and $cb \leq bc$, then
$$(bc)^n \leq b^n c^n < a^{n+1} c^n \leq (ac)^{n+1};$$

3. if $ca \leq ac$ and $bc \leq cb$, then
$$(bc)^n \leq c^n b^n < c^n a^{n+1} \leq c^{n+1} a^{n+1} \leq (ac)^{n+1};$$

4. if $ca \leq ac$ and $cb \leq bc$, then
$$(bc)^n \leq c^{n+1} b^n < c^{n+1} a^{n+1} \leq (ac)^{n+1}$$

for every n. A similar reasoning applies to the other pair. We conclude that the subdivision of S into the intervals I_λ is a compatible classification γ. It is readily seen that the factor semigroup $\bar{S} = S/\gamma$ is f. o. under the obvious ordering. $\bar{S}$ has no anomalous pair, for if $\bar{a}$, $\bar{b}$ were one, then $a \in \bar{a}$ and $b \in \bar{b}$ would form an anomalous pair in S. It follows at once that $\bar{S}$ is again cancellative. Thus, in view of the preceding section,

Theorem 6. (HION [3].) *Let S be a cancellative f. o. semigroup which is positively ordered and Archimedean. There exists an o-homomorphism of S into the additive semigroup of the non-*

negative real numbers such that two distinct elements of S have the same image if, and only if, they form an anomalous pair.

This result does not say anything about the structure of the intervals I_λ. HION [3] has shown that every f. o. set can occur as an I_λ.

5. Archimedean classes

We shall now consider the Archimedean classes of f. o. semigroups S. The elements $a, b \in S$ are called *Archimedean equivalent* (notation: $a \sim b$) if one of the following four possibilities holds for some positive integer n:

$$a \leq b \leq a^n, \quad b \leq a \leq b^n, \quad a^n \leq b \leq a, \quad b^n \leq a \leq b.$$

The first two cases may occur if $a, b \in P$, the other two if $a, b \in N$. It is routine to check that this is indeed an equivalence relation between the elements of S. Thus the elements of S fall into pair-wise disjoint *Archimedean classes*. The class containing a will be denoted by $\varkappa(a)$. Clearly, S is Archimedean if, and only if, $\varkappa(a) = \varkappa(b)$ for all $a, b \in P^*$ and for all $a, b \in N^*$.

Proposition 7. (HION [3].) *The subdivision of a f. o. semigroup S into Archimedean classes is a decomposition of S into pair-wise disjoint convex subsemigroups. Every decomposition of S into pair-wise disjoint convex subsemigroups can be refined to the decomposition into Archimedean classes.*

The convexity of Archimedean classes follows readily. That the classes are subsemigroups is likewise easy to see; for instance in the first case $a \leq b \leq a^n$ implies $a \leq ab \leq a^{n+1}$ and $a \leq ba \leq a^{n+1}$. Since $a \sim b$ implies that either the convex subsemigroup generated by a includes b or conversely, a and b cannot belong to disjoint convex subsemigroups of S.

The case of positively ordered semigroups deserves particular attention.

Theorem 8. (HION [3].) *Let S be a f. o. semigroup which is positively ordered. The Archimedean classes of S form a f. o. semigroup $\bar{S}$ under complex multiplication in which the product of two elements is equal to the greater of the two. The mapping*

$$a \to \varkappa(a)$$

*of S onto $\bar{S}$ is an o-homomorphism which may be characterized
as the minimal o-homomorphism of S onto f. o. semigroups in
which the product equals the least upper bound.*

The Archimedean classes are ordered in the obvious way
by setting $\varkappa(a) \leq \varkappa(b)$ whenever $a \leq b$. This is clearly a full
order. Since $\varkappa(b) = \varkappa(b^2)$, and $a \leq b$ implies $b \leq ab \leq b^2$, we
have $\varkappa(ab) = \varkappa(b)$ showing that $\varkappa(ab) = \max(\varkappa(a), (b))$. Therefore, the multiplication of Archimedean classes is well defined
by the rule $\varkappa(a) \varkappa(b) = \varkappa(ab)$, furthermore $a \to (a)$ is an
o-homomorphism. If $\eta\colon a \to a'$ is an o-homomorphism of S
onto a f. o. semigroup T in which $a'b' = \max(a', b')$, then the
operation in T is idempotent, and therefore η maps a convex
subsemigroup generated by one element upon a single element
of T. The stated characterization of the mapping $a \to \varkappa(a)$
follows now from Proposition 7. Q. E. D.

By an upper class in the semigroup $\bar{S}$ of Archimedean
classes of S we mean a subset U of $\bar{S}$ such that $\bar{u} \in U$, $\bar{v} \in \bar{S}$
and $\bar{u} \leq \bar{v}$ imply $\bar{v} \in U$. If U is an upper class, then the elements
$a \in S$ with $\varkappa(a) \in U$ form a convex prime ideal I of S—as one
readily verifies. This is a one-to-one correspondence between
the upper classes of $\bar{S}$ and the convex prime ideals of S. Therefore a necessary and sufficient condition that a f. o. semigroup
which is positively ordered be Archimedean is that it contain
no non-trivial convex prime ideal.

In case S is a naturally f. o. semigroup, it is easy to see
that the Archimedean classes of S are in a one-to-one correspondence with the principal (i. e. one element generated) convex
subgroups of the group G of quotients of S, namely, under the
(natural) correspondence $\varkappa(a) \to \{a\}_\square$ (see CONRAD [11]).

6. Ordinal sums

Let Λ be a f. o. set. To each $\lambda \in \Lambda$ we assign a f. o. semigroup S_λ such that for $\lambda \neq \mu$, S_λ and S_μ are disjoint. The set
union $S = \cup S_\lambda$ of all these S_λ will be made into a f. o. semigroup by the following definitions:

1. let the ordering relation and multiplication within each
S_λ be the same as originally given in S_λ;

2. if $a \in S_\lambda$, $b \in S_\mu$ and $\lambda < \mu$, then put $a < b$ and $ab = ba = b$.

One verifies easily that S is actually a f. o. semigroup, called the *ordinal sum* of the f. o. set $[S_\lambda \mid \lambda \in \varLambda]$ of f. o. semigroups S_λ. It is evident that S is positively ordered if, and only if, the same is true for every S_λ.

Proposition 9. (CLIFFORD [5].) *An ordinal sum $S = \cup S_\lambda$ is naturally ordered if, and only if, every S_λ is naturally ordered.*

If every S_λ is naturally ordered, then the definition of order and multiplication for elements of different S_λ's implies at once the statement for S. If, conversely, S is naturally ordered and $a < b$ $(a, b \in S_\lambda)$, then $b = ac = da$ for some $c, d \in S$. Since $c \notin S_\lambda$ would imply $ac = a$ or $ac = c$ according as $c < a$ or $a < c$, the proof is finished.

Call a f. o. semigroup *ordinally irreducible* if it cannot be expressed as an ordinal sum of two (or more) of its subsemigroups.

It follows at once that an Archimedean positively f. o. semigroup without identity is ordinally irreducible. The next theorem reduces the general case to ordinally irreducible semigroups.

Theorem 10. (KLEIN-BARMEN [2], CLIFFORD [5].) *Every positively f. o. semigroup can be uniquely represented as an ordinal sum of a f. o. set of ordinally irreducible, positively f. o. semigroups.*

Let S satisfy the hypotheses. By a *cut* in S will be meant a decomposition of S into two disjoint subsets L and U, called the lower and upper classes of the cut, such that $a \in L$, $x < a$ imply $x \in L$, and $b \in U$, $y > b$ imply $y \in U$. We call a cut (L, U) an *a-cut* if[7]

(i) L is a subsemigroup of S, and

(ii) $a \in L$ and $b \in U$ imply $ab = ba = b$.

The a-cuts can be f. o. by putting $(L_1, U_1) < (L_2, U_2)$ if L_1 is properly contained in L_2. An a-cut will be called a *β-cut* if it has an immediate successor a-cut in this ordering. Let $\varLambda$ denote

[7] Note that U is always an ideal in S. The cases when either L or U is void are also allowed.

the f. o. set of all β-cuts of S. To each $\lambda \in \Lambda$ we define S_λ as the intersection of the upper class U_λ of the β-cut (L_λ, U_λ) corresponding to λ with the lower class L'_λ of its immediate successor a-cut (L'_λ, U'_λ). Clearly, S_λ as an intersection of two convex subsemigroups is a convex subsemigroup of S. We claim that the S_λ are ordinally irreducible and S is their ordinal sum.

Supposing S_λ were the ordinal sum of its subsemigroups S_λ^1 and S_λ^2, we show that $(L_\lambda \cup S_\lambda^1, S_\lambda^2 \cup U'_\lambda)$ would be an a-cut between (L_λ, U_λ) and (L'_λ, U'_λ). The only point that requires verification is that (ii) is satisfied; the proof is straightforward by distinguishing the four possibilities as to inclusions of a, b. Since the S_λ satisfy the hypotheses of an ordinal sum, we infer that S contains the ordinal sum of the $[S_\lambda \mid \lambda \in \Lambda]$. Thus it remains only to prove that every element a of S belongs to some S_λ.

To an $a \in S$ let us consider the union L_0 of all lower classes of a-cuts (L, U) with $a \notin L$ (including the void set) and the intersection L_1 of all L's with $a \in L$. Then $(L_0, S \setminus L_0)$ and $(L_1, S \setminus L_1)$ are a-cuts in S, and it follows directly that no a-cut lies between them. Therefore $(L_0, S \setminus L_0)$ is a β-cut, and if S_λ corresponds to this β-cut, then $a \in S_\lambda$.

To prove the uniqueness, assume that S is the ordinal sum of the ordinally irreducible f. o. semigroups T_μ ($\mu \in M$). It is immediate that

$$L_\varkappa = \bigcup_{\mu < \varkappa} T_\mu, \quad U_\varkappa = \bigcup_{\mu \geq \varkappa} T_\mu$$

define an a-cut $(L_\varkappa, U_\varkappa)$ and

$$L'_\varkappa = \bigcup_{\mu \leq \varkappa} T_\mu, \quad U'_\varkappa = \bigcup_{\mu > \varkappa} T_\mu$$

define an a-cut $(L'_\varkappa, U'_\varkappa)$. Because of the ordinal irreducibility of $T_\varkappa$ it is impossible to intercalate between them a third a-cut, thus $(L_\varkappa, U_\varkappa)$ is a β-cut of S with $(L'_\varkappa, U'_\varkappa)$ as immediate successor a-cut. Consequently, for some $\lambda \in \Lambda$, $S_\lambda = U_\varkappa \cap L'_\varkappa = = T_\varkappa$. Since the union of all $T_\varkappa$ exhausts S, every S_λ occurs as a $T_\varkappa$. This completes the proof of Theorem 10.

It is natural to ask for conditions which ensure that a positively f. o. semigroup S is the ordinal sum of ordinally irreducible semigroups of some important type. As an example of such results we prove:

Proposition 11. (CONRAD [11].) *For a positively f. o. semigroup S to be an ordinal sum of cancellative, ordinally irreducible f. o. semigroups the following condition is necessary and sufficient*:

$$ab = ac \ (or \ ba = ca) \ implies \ b = c \ or \ ab = a \ (ba = a).$$

It is readily seen that an ordinal sum of semigroups of the stated kind fulfils this condition. Consequently, let us assume that S is a positively f. o. semigroup that satisfies the condition.

First of all we show that a product ab is idempotent only if one of the factors is idempotent and $ab = \max(a, b)$. If a is not idempotent, then $a < a^2 \leqq aba$. Therefore $abab = ab$ implies by the supposed condition $ab = b$. Since $a \leqq ab$, we have $\max(a, b) = b$.

Now we define an equivalence relation τ in S. Let $a \tau b$ mean that either a, b are equal or they are not idempotent and

$$\min(ab, ba) > \max(a, b).$$

Then τ is obviously symmetric and reflexive. If $a \tau b$ and $b \tau c$, then also $a \tau c$. For if, say, neither of a, b, c is idempotent and if say $ac = a$, then $acb = ab$ would imply $cb = b$ or $ab = a$, contrary to $b \tau c$ or $a \tau b$. Thus τ is an equivalence relation, and so S splits into disjoint classes under τ.

The classes S_λ under τ are subsemigroups. We show that if a, b are not idempotent and $a \tau b$, then $a \tau ab$. The assumption $a^2 b = ab$ would lead to $ab = b$ or to $ab = a$, while $aba = ab$ to $ba = b$ or to $ab = a$. Either case is impossible, thus $a \tau ab$.

The classes S_λ are cancellative, for if a, b, c belong to the same class and $ab = ac$, then $ab > a$ implies $b = c$, and symmetrically for the other side. That the S_λ are ordinally irreducible is immediate from the definitions. It is also readily seen that the S_λ are convex, hence we may define $S_\lambda < S_\mu$ if for representatives $a \in S_\lambda$, $b \in S_\mu$ the relation $a < b$ holds.

Since this is a full order in the set of classes S_λ, it suffices to establish that S is the ordinal sum of these S_λ. If $a < b$ and a, b belong to different classes, then either $ab = b$ or $ba = b$. We have to prove that both hold. Suppose $ab = b$. If b is idempotent, then $b \leq ba \leq b^2 = b$, while if it is not, then $bab = b^2$ implies $ba = b$. In either event $ba = b$, and this completes the proof of all parts of the theorem.

7. Completion of fully ordered semigroups

If we start with a p. o. semigroup S and form its Dedekind—MacNeille completion $S^\sharp$ by cuts in the usual fashion, then it turns out that there are in general several ways of extending the operation of S to $S^\sharp$ even if S is f. o. and the multiplication is continuous in the order topology of S. It is instructive to investigate this new phenomenon in some detail.

In this section S will denote a f. o. semigroup. S is called *lower semi-continuous* if

$$c < ab \quad (a, b, c \in S)$$

implies the existence of neighbourhoods[8] U_a of a and U_b of b such that

$$c < xy \quad \text{for all } x \in U_a, \, y \in U_b.$$

Upper semi-continuity is defined dually. It is readily checked that continuity of multiplication is equivalent to lower and upper semi-continuity of S.

Lemma A. (CLIFFORD [8].) *S is lower semi-continuous if, and only if, it fulfils the following condition*: if A, B are subsets of S for which sup A and sup B exist,[9] then sup AB exists and

$$\sup AB = \sup A \sup B.$$

Let S be lower semi-continuous and $a = \sup A$, $b = \sup B$. Then ab is an upper bound for AB. Let $c \in S$ be such that

---

[8] The neighbourhoods U_a are open intervals containing a. The situation described above will be written for brevity as $c < U_a U_b$.

[9] Recall that a is the supremum of the set A if $U(A) = U(a)$.

$c < ab$. Then there exist neighbourhoods U_a of a and U_b of b such that $c < U_a U_b$. Since neither $A \cap U_a$ nor $B \cap U_b$ is empty, we have $c < a'b'$ for some $a' \in A$, $b' \in B$. Thus $ab = = \sup AB$.

Conversely, suppose the stated condition and let $c < ab$ $(a, b, c \in S)$. If neither $S_a = [x \in S \mid x < a]$ nor $S_b = [x \in S \mid x < b]$ has a greatest element, then $a = \sup S_a$ and $b = = \sup S_b$, whence $ab = \sup S_a S_b$. Because of $c < ab$ there exist $a_1 \in S_a$, $b_1 \in S_b$ such that $c < a_1 b_1$. Now $U_a = [x \in S \mid x > a_1]$ and $U_b = [x \in S \mid x > b_1]$ are open sets containing a and b, respectively, such that $c < U_a U_b$. — If S_a has a greatest element a', then we argue with a rather than S_a and use $U_a = [x \in S \mid x > a']$. Similarly for S_b. Q. E. D.

We proceed to define (on the pattern of Section **10** of Chapter V) the closure operation

$$X \to X^{\#} = L(U(X))$$

for non-void u-bounded subsets X in a lower semi-continuous f. o. semigroup S. The set $S^{\#}$ of all "closed" subsets is f. o. by inclusion and the correspondence

$$a \to a^{\#} = L(a) \quad (a \in S)$$

embeds S in $S^{\#}$; it is actually order-preserving. In order to make $S^{\#}$ into a semigroup, define for A, $B \in S^{\#}$ the product[10] $A \cdot B$ as

$$A \cdot B = (AB)^{\#}.$$

The associativity of this product will follow from the identity

(1) $(X^{\#} Y^{\#})^{\#} = (XY)^{\#}$ for $X, Y \subseteq S$.

It is clearly sufficient to verify $U(X^{\#} Y^{\#}) = U(XY)$. Assume the contrary: w is an upper bound for XY, but not for $X^{\#} Y^{\#}$. Then $w < x'y'$ for some $x' \in X^{\#}$, $y' \in Y^{\#}$. Observe that if $x' \leq x$ for no $x \in X$, then $x' = \sup X$. From Lemma A we

---

[10] We have to distinguish between complex multiplication AB of the sets A, B and their product $A \cdot B$ in $S^{\#}$.

conclude that either $w < \sup Xy$ for some $y \in Y$ or $w < < \sup XY$, proving (1). Since

$$(A \cdot B) \cdot C = ((AB)^{\#} C)^{\#} = ((AB)C)^{\#} = (A(BC))^{\#} =$$
$$= (A(BC)^{\#})^{\#} = A \cdot (B \cdot C)$$

and

$$A \subseteq B \text{ implies } A \cdot C = (AC)^{\#} \subseteq (BC)^{\#} = B \cdot C$$

for all $A, B, C \in S^{\#}$, $S^{\#}$ is a f. o. semigroup in which S is a subsemigroup. The (conditional) completeness of $S^{\#}$ is readily verified.

$S^{\#}$ is lower semi-continuous again. For, if Γ, Δ are subsets of $S^{\#}$ such that $\sup \Gamma$ and $\sup \Delta$ exist in $S^{\#}$, then

$$\sup \Gamma \cdot \sup \Delta \geq \sup (\Gamma \cdot \Delta)$$

and

$$\sup \Gamma \cdot \sup \Delta = (\vee C)^{\#} \cdot (\vee D)^{\#} = ((\vee C)^{\#} (\vee D)^{\#})^{\#} =$$
$$= ((\vee C)(\vee D))^{\#} = (\vee CD)^{\#} \leq (\vee (CD)^{\#})^{\#} = (\vee (C \cdot D))^{\#} =$$
$$= \sup (\Gamma \cdot \Delta)$$

with $C \in \Gamma$, $D \in \Delta$. We have thus proved

Theorem 12. (KRISHNAN [1], CLIFFORD [8].) *Every lower semi-continuous f. o. semigroup can be embedded in a complete lower semi-continuous f. o. semigroup.*

In order to obtain a unicity statement, let us call a complete f. o. set T a *normal completion* of the f. o. set S if T contains S such that every element of T is the l. u. b. of some subset of S and also the g. l. b. of some subset of S. Normal completions can be described in terms of l. u. b. only as follows:

Lemma B. (CLIFFORD [8].) *A complete f. o. set T is a normal completion of its subset S if, and only if,*

(i) *every element of T is the l. u. b. of a subset of S,*

(ii) *if $a \in S$ is the l. u. b. of a subset A of S relative to S, then a is the l. u. b. of A relative to T,*

(iii) *S contains the greatest element of T if such exists.*

Let T be a normal completion of S. To establish (ii), let $a \in S$ be the l. u. b. of A ($\subseteq S$) relative to S and a the l. u. b.

of A relative to T. Obviously $a \leq a$. Strict inequality $a < a$ is impossible, for in this event no element of S lies between a and a, and so there exists no subset of S for which a is the g. l. b. (iii) holds, because the greatest element must be a g. l. b.

Conversely, assume (i)—(iii). Pick $a \in T \setminus S$ arbitrarily, and consider $S_a = [x \in S \mid x \geq a]$. By (iii), S_a is not empty. Let β be the g. l. b. of S_a in T. Then $a \leq \beta$, and it is enough to prove that $a < \beta$ is impossible. $a < \beta$ and $\beta \notin S$ would contradict (i), for between a and β there is no element of S. Further, $a < \beta$ and $\beta \in S$ would imply that β is the l. u. b. of $[x \in S \mid x \leq a]$ relative to S and a is that relative to T, contrary to (ii). Q. E. D.

Now we are in a position to show that $S^{\sharp}$ as defined above is a normal completion of S, that is, $S^{\sharp}$ satisfies (i)—(iii) of the preceding lemma. If $A \in S^{\sharp}$, then A, considered now as a subset of S, is not void. Let B be an upper bound of A in $S^{\sharp}$. Then every $x \in A$ satisfies $x^{\sharp} \subseteq B^{\sharp}$, $x \in B$, and so $A \subseteq B$. Thus A is the l. u. b. for A in $S^{\sharp}$, and (i) holds. Next assume that $a \in S$ is the l. u. b. of a subset X of S relative to S, and let A be the l. u. b. for X relative to $S^{\sharp}$. Then $X \subset A$ and $X^{\sharp} \subseteq A^{\sharp} = A$. From $a \in S$, $a \in X^{\sharp}$ we get $a \in A$, $a^{\sharp} = A$, proving (ii). Since $S^{\sharp}$ has a greatest element if, and only if, S is bounded from above, i. e. S has one, condition (iii) is fulfilled.

We intend to prove that every lower semi-continuous normal completion T of S is o-isomorphic to $S^{\sharp}$ over S. We define

$$f(a) = S_a = [x \in S \mid x \leq a] \quad \text{for } a \in T.$$

Then $f(a) = a^{\sharp}$ for all $a \in S$. It follows that $f(a)^{\sharp} = f(a)$, for $a \in f(a)^{\sharp} \setminus f(a)$ would lead to a contradiction to (ii). Thus $f(a)$ is a single-valued and, obviously, order-preserving mapping from T into $S^{\sharp}$. Moreover, it is one-to-one, for a is the l. u. b. of S_a relative to T. To see that it is exhaustive, let $A \in S^{\sharp}$ and a the l. u. b. of A relative to T. Then $A \subset S_a$ and if a $b \in S_a$ were not in A, then it would be an upper bound of A, and so $b \geq a$, $b = a$ and $b \in A$, a contradiction. Hence $A = S_\alpha$ and $f(a) = A$. Finally, we prove that $f(a\beta) = = f(a) \cdot f(\beta)$, holds, or, in an equivalent form, $f^{-1}(A \cdot B) = = f^{-1}(A)f^{-1}(B)$ for A, $B \in S^{\sharp}$. We have

$$f^{-1}(A) f^{-1}(B) = \sup A \sup B = \sup AB =$$

$$= f^{-1}((AB)^{\sharp}) = f^{-1}(A \cdot B),$$

as claimed. This completes the proof of

Theorem 13. (CLIFFORD [8].) *A lower semi-continuous f. o. semigroup S has a lower semi-continuous normal completion which is unique up to o-isomorphisms over S.*

If we assume that S is continuous, then it has an upper semi-continuous normal completion as well. We are naturally interested to know whether this must be isomorphic to the lower semi-continuous normal completion. It is easy to see that this is not the case as shown by the following example, due to CLIFFORD [8]. Let S be the real interval $[0, 1]$ with $\frac{1}{2}$ removed and define $ax = xa = 0$ for all $a \in [0, \frac{1}{2})$ and $x \in S$, while $bc = \frac{1}{4}$ for all $b, c \in (\frac{1}{2}, 1]$. Then the Dedekind completion S_0 of S is $[0, 1] = S \cup \frac{1}{2}$. Now if we define

$$a \cdot \frac{1}{2} = \frac{1}{2} \cdot a = 0, \quad \frac{1}{2} \cdot \frac{1}{2} = a, \quad \frac{1}{2} \cdot x = x \cdot \frac{1}{2} = f(x)$$

$$\text{for } a \in [0, \tfrac{1}{2}), \, x \in (\tfrac{1}{2}, 1]$$

where a is an arbitrary real number between 0 and $\frac{1}{4}$, and $f(x)$ is an arbitrary monotone non-decreasing mapping from $(\frac{1}{2}, 1]$ into $[a, \frac{1}{4}]$, then we obtain a normal completion $S_0(a, f)$ of S. The lower and upper semi-continuous normal completions correspond to the cases $a = 0$, $f(x) = 0$, and $a = \frac{1}{4}$, $f(x) = \frac{1}{4}$, respectively.

The normal completions of a continuous f. o. semigroup S can be p. o. Assume, as may be done, that all normal completions are defined on the same set, namely, on the Dedekind completion T of S, regarded as a f. o. set. Then we define $T_1 \leq T_2$ to mean that the product of no a, β $(\in T)$ taken in T_1 is greater than their product in T_2. With this definition, the lower semi-continuous normal completion is the least of all, and the upper one is the greatest—a fact which is readily checked from the definitions.

Theorem 14. (CLIFFORD [8].) *A continuous f. o. semigroup S has only one normal completion exactly in case its lower and upper semi-continuous normal completions coincide. This happens if, and only if, the product of two cuts of S is again a cut of S.*

Here we mean by a *cut* a pair (L, U) of non-void subsets L, U of S such that $x \leq y$ for every $x \in L$ and $y \in U$, and there is at most one $a \in S$ satisfying $x \leq a \leq y$ for all $x \in L$, $y \in U$.[11] The product of the cuts (L_1, U_1) and (L_2, U_2) is defined as $(L_1 L_2, U_1 U_2)$ (which need not be a cut).

To prove the second part of the theorem, suppose that T is the only normal completion of S and (L_1, U_1), (L_2, U_2) are two cuts in S. Write $a = \sup L_1$ and $\beta = \sup L_2$. Then we also have $a = \inf U_1$ and $\beta = \inf U_2$. Consequently, by Lemma A and its dual, $\sup L_1 L_2 = a\beta = \inf U_1 U_2$, whence $(L_1 L_2, U_1 U_2)$ is again a cut. Conversely, let S satisfy the stated condition and let T be the Dedekind completion of S, considered as a f. o. set. Any $a \in T$ defines a cut (L_a, U_a) where $L_a = [x \in S \mid x \leq a]$, $U_a = [x \in S \mid x \geq a]$. From hypothesis we get that $(L_a L_\beta, U_a U_\beta)$ is again a cut, i. e. $\sup L_a L_\beta = \inf U_a U_\beta$. Here the left member is nothing else than the product of a and β in the lower semi-continuous normal completion of S, while the right member is the same for the upper one. We arrive at the coincidence of the lower and upper semi-continuous normal completions, as we wished to show.[12]

CLIFFORD [8] investigates the case of commutative naturally f. o. semigroups and gives criteria under which they are complete or normally embeddable in complete naturally f. o. semigroups.

KRISHNAN [1], [2] discusses the completions of p. o. semigroups and their interrelations.

8. On a class of fully ordered groupoids

Having discussed f. o. semigroups we may try to drop the associativity assumption and to obtain results on f. o. groupoids

---

[11] It is easy to see that then $\sup L = \inf U = a$.

[12] Even an Abelian f. o. group may, if its order is non-Archimedean, have different lower and upper semi-continuous normal completions.

in general. However, such an endeavour is obviously hopeless
unless some restrictive assumptions are made on the groupoids
considered. We have selected a class of f. o. groupoids which
owes its interest, in the first place, to the part it plays in the
study of mean values and more generally, in functional
equations.[13]

A *mean groupoid* is a set M with the properties:

   (i) M is a strict f. o. groupoid;

  (ii) every element of M is idempotent;

 (iii) M satisfies the *bisymmetric* law:[14]

$$(ab)(cd) = (ac)(bd) \quad \text{for all } a, b, c, d \in M;$$

 (iv) M is *Archimedean* in the sense that if $a < c < b$, then
there is a natural number n such that multiplying a by b
n times on the left (right), we have $bb \ldots ba > c$ $(abb \ldots b >$
$> c)$; and dually.[15]

Some immediate consequences may be mentioned:

A) Multiplication is internal: $a < b$ implies $a < ab < b$
and $a < ba < b$. This follows from strict monotony and idempotency. — Hence M is dense-in-itself.

B) The *autodistributive* laws hold:

$$a(bc) = (ab)(ac) \quad \text{and} \quad (bc)a = (ba)(ca)$$

for all $a, b, c \in M$. This is immediate from $a = aa$ and bisymmetry.

C) The operation is continuous in the open-interval topology. First we show that if $c < ab$ then $c < U_a b$ for some neigh-

---

[13] For the theory of mean values see J. AczÉL, *Bull. Amer. Math.
Soc.*, **54** (1948), 392—400. The crucial axiom (iii) has been used formerly
by him [*Norske Vid. Selsk. Forh.*, **19** (1946), 83—86] and in the case
of quasigroups by K. TOYODA, *Tohoku Math. Journ.*, **46** (1940), 239—
251. The results of this section, with M an interval of real numbers,
are due to AczÉL; most of the proofs go back to him. Cf. also J. AczÉL,
Vorlesungen über Funktionalgleichungen und ihre Anwendungen (Basel—
Berlin, 1961).

[14] This is also known as the entropic law; cf. I. M. H. ETHERINGTON,
Proc. Roy. Soc. Edinb., (A) **62** (1949), 442—453, or A. SADE, *Rev. Fac.
Sci. Univ. Istanbul*, (A) **22** (1957), 151—184.

[15] Parentheses can be removed without danger of ambiguity, since
clearly $bb \ldots ba = \{b[\ldots(ba)]\}$.

bourhood U_a of a. Only the case $c \in [a, b]$ (or $c \in [b, a]$) is not trivial. Then we have by the Archimedean property, for a fixed $x < a$, the inequality

$$c < (xb)\,(ab)\,(ab) \ldots (ab) = (xaa \ldots a)b$$

with sufficiently large number of factors. Thus $c < U_a\, b$ for $U_a = [y \in M \mid y > xa \ldots a]$. Now we prove that if $c < ab$ then $c < U_a\, U_b$ for certain neighbourhoods U_a of a and U_b of b. We know that $c < a_1\, b < ab$ for some $a_1 < a$, and similarly $c < a_1\, b_1 < a_1\, b$ for some $b_1 < b$. This establishes lower semi-continuity. Combining this with its dual, the assertion follows.

The discussion of mean groupoids begins with the commutative case.

Theorem 15. (FUCHS [6].) *Every commutative mean groupoid M has an order-preserving one-to-one mapping f into the real numbers such that*

$$(1) \qquad f(xy) = \tfrac{1}{2}[f(x) + f(y)] \quad \text{for all } x, y \in M.$$

This mapping f is unique up to linear transformations.

Select two distinct elements of M, say, $a < b$. We shall first define f on the subsystem M' generated by a, b with proper dyadic fractions as values. It is more convenient to construct the inverse f^{-1} of f. Put

$$f^{-1}(0) = a \quad \text{and} \quad f^{-1}(1) = b.$$

Let $k/2^n$ be a proper dyadic fraction and write $k = 2q + r$ with $r = 0$ or 1 (k, q are non-negative integers). Supposing f^{-1} to be defined for proper dyadic fractions with denominators less than 2^n, we set

$$f^{-1}\left(\frac{k}{2^n}\right) = f^{-1}\left(\frac{q}{2^{n-1}}\right) f^{-1}\left(\frac{q+r}{2^{n-1}}\right).$$

Evidently, f^{-1} is strictly increasing. In order to show that it satisfies

$$f^{-1}(\xi)\, f^{-1}(\eta) = f^{-1}\left(\frac{\xi + \eta}{2}\right)$$

for all dyadic fractions ξ, η in the interval $[0, 1]$, we proceed by induction. Write $\xi = (2q + r) / 2^n$ and $\eta = (2q' + r') / 2^n$ with r, r' equal to 0 or 1. Then

$$f^{-1}(\xi) \, f^{-1}(\eta) = \left[f^{-1}\left(\frac{q}{2^{n-1}}\right) f^{-1}\left(\frac{q+r}{2^{n-1}}\right) \right] \cdot \left[f^{-1}\left(\frac{q'+r'}{2^{n-1}}\right) f^{-1}\left(\frac{q'}{2^{n-1}}\right) \right] =$$

$$= \left[f^{-1}\left(\frac{q}{2^{n-1}}\right) f^{-1}\left(\frac{q'+r'}{2^{n-1}}\right) \right] \cdot \left[f^{-1}\left(\frac{q+r}{2^{n-1}}\right) f^{-1}\left(\frac{q'}{2^{n-1}}\right) \right] =$$

$$= f^{-1}\left(\frac{q+q'+r'}{2^n}\right) f^{-1}\left(\frac{q+q'+r}{2^n}\right) =$$

$$= f^{-1}\left(\frac{2q + 2q' + r + r'}{2^{n+1}}\right) = f^{-1}\left(\frac{\xi + \eta}{2}\right)$$

(we have used the definition, commutativity, bisymmetry, induction hypothesis and again the definition). Therefore f is an isotone and one-to-one mapping from M' onto the dyadic fractions in $[0, 1]$ satisfying (1).

To extend f to the interval $[a, b]$, let $a < c < b$. The element c defines two sets L_c, U_c of dyadic fractions in $[0, 1]$ if we let $\xi \in L_c$ whenever $f(x) = \xi$ for some $x \in M'$, $x < c$, and $\eta \in U_c$ whenever $f(y) = \eta$ for some $y \in M'$, $c < y$. Obviously, (L_c, U_c) is a Dedekind cut in $[0, 1]$. If ζ is the real number which it determines, then we define $f(c) = \zeta$. The assumption $f(c) = f(d) = \zeta$, $c < d$ leads to a contradiction; for if this holds, then M' has no element between c and d. and if $c < add \ldots d$ with n factors d, then $x < ayy \ldots y$ for all $f(x) \in L_c$, $f(y) \in U_d = U_c$. We can apply (1) to the elements of M', and therefore

$$f(x) < f(ayy \ldots y) = \frac{2^n - 1}{2^n} f(y) \quad \text{for all} \quad f(x) \in L_c, f(y) \in U_c,$$

a contradiction. Thus f is a one-to-one and clearly isotone function from the interval $[a, b]$ of M into $[0, 1]$. It satisfies (1) which follows readily from C).

Finally, we extend f to the whole of M. Let e. g. $b < c$. The real number ζ corresponding to c may be calculated without any difficulty on the basis of $a < c' = au \ldots ac < b$ and the value of $f(c')$. It is now easy to verify that the extended f has the desired properties.

If f is a mapping with these properties, then so is the mapping g defined by $g(x) = \lambda f(x) + \mu$ where $\lambda > 0$, μ are real numbers. No other function g has the required properties, for the real function gf^{-1} satisfies the functional equation

$$gf^{-1}[\tfrac{1}{2}(\xi + \eta)] = \tfrac{1}{2}[gf^{-1}(\xi) + gf^{-1}(\eta)].$$

The only monotone solutions of this equation are the linear functions $\lambda \xi + \mu$. This finishes the proof.[16]

We now turn to the non-commutative case. Here the above procedure breaks down and in the reduction to the commutative case completeness seems to be required. In order to embed a mean groupoid M in a complete mean groupoid M^*, we assume

(v) the product of every pair of cuts in M is again a cut of M.

Then the Dedekind completion of M can be made into a mean groupoid by the method of the foregoing section.

Theorem 16. (FUCHS [6].) *A mean groupoid M satisfying (v) is o-isomorphic to a subset of real numbers endowed with the operation of forming weighted arithmetic means. More explicitly, to M there exist an order-preserving one-to-one mapping f from M into the real line and a real number $\lambda\,(0 < \lambda < 1)$ such that*

(2) $\qquad f(xy) = \lambda f(x) + (1 - \lambda)f(y)$ *for all* $x, y \in M$.

In view of our previous observation, it suffices to consider the case when M is complete. This assumption guarantees that, given $t, x, y \in M$, there exists a unique $z \in M$ between x and y satisfying

(3) $\qquad\qquad (tz)\,(zt) = (tx)\,(yt).$

---

[16] In the light of this theorem the terminology of "Archimedean" in our present sense may be reformulated: if $a < \gamma < \beta$, then there exists an n such that $\gamma < 2^{-n}[a + (2^n - 1)\beta]$, that is, $2^n\,(\beta - \gamma) > \beta - a$ which is actually nothing else but the well-known Archimedean axiom.

(This is essentially BOLZANO's theorem which holds because of completeness and continuity.) Keeping t fixed, we consider z as a function of x and y, and write[17]

$$z = x \mathsf{T} y.$$

We intend to show that T is a commutative mean operation on M. It is evidently commutative, idempotent and strictly monotone. Since

$$[t(xy)]\,[(yx)t] = [(tx)\,(ty)]\,[(yt)\,(xt)] = [(tx)\,(yt)]\,[(ty)\,(xt)] =$$
$$= (tx)\,(yt) = (tz)\,(zt),$$

we obtain $x \mathsf{T} y \geq \min{(xy, yx)}$ and therefore $(x \mathsf{T} y) \mathsf{T} y \geq$ $\geq \min{(xyy, yxy, yyx)}$ and, by induction, the Archimedean property of T follows.

In the proof of bisymmetry we use the identity (of ACZÉL)

$$(4) \qquad (x \mathsf{T} y)\,(u \mathsf{T} v) = xu \mathsf{T} yv \quad \text{for all } x, y, u, v \in M.$$

This is valid, because with the notation $z = x \mathsf{T} y$, $w = u \mathsf{T} v$ we have

$$[t(zw)]\,[(zw)t] = [(tz)\,(zt)]\,[(tw)\,(wt)] = [(tx)\,(yt)]\,[(tu)\,(vt)] =$$
$$= [t(xu)]\,[(yv)t].$$

Note that (4) implies distributivity

$$x\,(u \mathsf{T} v) = xu \mathsf{T} xv, \quad (u \mathsf{T} v)x = ux \mathsf{T} vx,$$

since $x \mathsf{T} x = x$. Now in $a = (x \mathsf{T} y) \mathsf{T} (u \mathsf{T} v)$ the elements y and u can be interchanged, for

---

[17] Actually $z = x \mathsf{T} y$ is independent of t, for if u is another element of M and

$$(tz)\,(zt) = (tx)\,(yt), \quad (uw)\,(wu) = (ux)\,(yu),$$

then

$$u\,[(tx)\,(yt)]\,u = [(utu)\,(uxu)]\,[(uyu)\,(utu)] = [(uux)\,(tuu)]\,[(uyu)\,(utu)] =$$
$$= \{u\,[(ux)\,(yu)]\}\,\{(tuu)\,(utu)\} = \{u\,[(uw)\,(wu)]\}\,\{(tuu)\,(utu)\}.$$

Thus in $u\,[(tx)\,(yt)]u$ both elements x, y can be replaced by w, i. e.

$$u\,[(tz)\,(zt)]\,u = u\,[(tx)\,(yt)]u = u\,[(tw)\,(wt)]u$$

whence $w = z$.

$$(ta)(at) = [t(x \mathsf{T} y)][(u \mathsf{T} v)t] = (ty \mathsf{T} tx)(ut \mathsf{T} vt) =$$
$$= (ty)(ut) \mathsf{T} (tx)(vt)$$

by (4), and the original operation is bisymmetric.

We have proved that under the operation T, M is a commutative mean groupoid. From Theorem 15 we infer the existence of an order-preserving, one-to-one mapping f of M into the real line such that

$$f(x \mathsf{T} y) = \tfrac{1}{2}[f(x) + f(y)] \quad \text{for} \ x, y \in M.$$

With the notation $f(x) = \xi$, $f(y) = \eta$, $f(u) = \varrho$, $f(v) = \sigma$, (4) takes the form

$$f[f^{-1}(\tfrac{1}{2}(\xi + \eta)) \cdot f^{-1}(\tfrac{1}{2}(\varrho + \sigma))] =$$
$$= \tfrac{1}{2}(f[f^{-1}(\xi)f^{-1}(\varrho)] + f[f^{-1}(\eta)f^{-1}(\sigma)]).$$

This shows that the real function $F(\xi, \eta) = f[f^{-1}(\xi)f^{-1}(\eta)]$ satisfies JENSEN's functional equation

$$F[\tfrac{1}{2}(\xi + \eta), \tfrac{1}{2}(\varrho + \sigma)] = \tfrac{1}{2}[F(\xi, \varrho) + F(\eta, \sigma)]$$

whose only monotone solutions are the linear functions. Therefore $f^{-1}(\xi)f^{-1}(\eta) = f^{-1}(\lambda \xi + \mu \eta + \nu)$ for fixed real $\lambda > 0$, $\mu > 0$, ν. The cases $\xi = \eta = 0$ and $\xi = \eta = 1$ yield $\nu = 0$ and $\lambda + \mu = 1$, completing the proof.[18]

M. Hosszú[19] observed that (ii) and (iii) may be replaced by the postulates that multiplication is internal and the autodistributive laws hold. In fact, (ii) follows at once if we put $a = b$ in the first autodistributive law. To prove the bisymmetric law, we first embed the groupoid in a complete one which again obeys the autodistributive laws. Then we define v^* as the solution of the equation

$$(xy)(uv) = (xu)(yv^*)$$

for fixed x, y, u and for v varying between y and u. Since $(xu)(yy) = (xu)y = (xy)(uy) = (xu)(yy^*)$, we have $y^* = y$ and similarly $u^* = u$. Furthermore,

$$(vw)^* = v^* w^*,$$

---

[18] From the proof it turns out that the only point where we have made use of the additional assumption (v) is the conclusion that (3) has always a solution z. Hence it would be equally well to suppose this instead of (v).

[19] *Publicationes Math. Debrecen*, **6** (1959), 1—6.

for

$$(xu) [y(vw)^*] = (xy) [u(vw)] = [(xy) (uv)] [(xy) (uw)] =$$
$$= [(xu) (yv^*)] \cdot [(xu) (yw^*)] = (xu) [y(v^* w^*)].$$

We see that $v^* = v$ for the elements v of the subgroupoid M' generated by u, y. The elements of M' lie everywhere dense between y and u, therefore continuity implies $v^* = v$ for all v between y and u. A similar reasoning applies in the other cases, e. g. if u lies between y and v.

J. Aczél[20] pointed out the case when the hypothesis of idempotency is dropped. Then one obtains a function f from the groupoid M into the real numbers which is one-to-one, strictly isotone and satisfies

$$f(xy) = \lambda f(x) + \mu f(y) + \nu$$

for fixed real numbers $\lambda > 0$, $\mu > 0$, ν and for all x, $y \in M$.

If strict monotony is replaced by

$$a < b \text{ implies } ac < bc \text{ and } ca > cb,$$

then the operation is external: neither ab nor ba lies between a and b. Theorem 16 remains valid with $\lambda > 1$.

[20] See his paper in the *Bulletin* cited in footnote [13].

CHAPTER XII

LATTICE-ORDERED SEMIGROUPS

1. Residuals

Let H be a p. o. groupoid and suppose that it has the following property: for all $a, b \in H$ there exists an element $a : b$ of H such that

$$x \leq a : b \text{ is equivalent to } xb \leq a.$$

Then we say that $a : b$ is the *right-residual* of a by b and H is a *right-residuated groupoid*. The *left-residual* of a by b is defined similarly as an element $a :: b$ such that

$$x \leq a :: b \text{ if, and only if, } bx \leq a.$$

A p. o. groupoid that is both right- and left-residuated is called *residuated*. Clearly, $a : b$ and $a :: b$ are, if they exist, uniquely determined by a and b.

Residuals generalize the concept of the ideal quotient in ring theory, and are especially important in l. o. semigroups. Many properties of the ideal structure of a ring carry over to such semigroups.

We begin by listing some elementary properties of residuals which are valid quite generally in a p. o. groupoid whenever the required residuals exist. Most of the statements follow directly from the definitions.[1]

　　i) $a \leq b$ implies $a : c \leq b : c$.

　　ii) $a \leq b$ implies $c : b \leq c : a$.

　　iii) The relations $bc \leq a$, $b \leq a : c$ and $c \leq a :: b$ are equivalent.

　　iv) $ab : b \geq a$ and $ba :: b \geq a$.

　　v) $a :: (a : b) \geq b$ and $a : (a :: b) \geq b$.

　　vi) $a : [a :: (a : b)] = a : b$ and $a :: [a : (a :: b)] = a :: b$.

　　vii) If H is negatively ordered, then $a \leq a : b$.

　　viii) Next we verify the important lemmas :

[1] See CERTAINE [1], DILWORTH [1], WARD—DILWORTH [1] and BIRKHOFF [3].

Lemma A. *If* $\bigvee a_\alpha$ $(a_\alpha \in H)$ *exists in the right-residuated groupoid* H, *then, for all* $b \in H$, $\bigvee(a_\alpha b)$ *also exists and*

$$\bigvee(a_\alpha b) = (\bigvee a_\alpha)b.$$

We have clearly $a_\alpha b \leq (\bigvee a_\alpha)b$. Assume that $a_\alpha b \leq x$ holds for all α. Then $a_\alpha \leq x : b$, $\bigvee a_\alpha \leq x : b$, and so $(\bigvee a_\alpha)b \leq x$. This proves that $(\bigvee a_\alpha)b$ is the l. u. b. of the $a_\alpha b$.

Lemma B. *If* H *is a right-residuated groupoid and* $\bigwedge a_\alpha$ $(a_\alpha \in H)$ *exists in* H, *then so does* $\bigwedge(a_\alpha : b)$ *for all* $b \in H$ *and*

$$\bigwedge(a_\alpha : b) = (\bigwedge a_\alpha) : b.$$

In fact, $c = (\bigwedge a_\alpha):b$ satisfies $cb \leq \bigwedge a_\alpha \leq a_\alpha$, that is to say, $c \leq a_\alpha : b$ for all α. If $x \leq a_\alpha : b$ for all α, then $xb \leq a_\alpha$, $xb \leq \bigwedge a_\alpha$, $x \leq (\bigwedge a_\alpha) : b = c$; in other words, c is the g. l. b. of the $a_\alpha : b$.

Lemma C. *If* $\bigvee b_\alpha (b_\alpha \in H)$ *exists in the residuated groupoid* H, *then so does* $\bigwedge(a : b_\alpha)$ *for all* $a \in H$ *and*

$$\bigwedge(a : b_\alpha) = a : (\bigvee b_\alpha).$$

The same holds for the left-residual.

The inequality $a : (\bigvee b_\alpha) \leq a : b_\alpha$ holds by ii). Let $x \leq$ $\leq a : b_\alpha$ for all α. Then $xb_\alpha \leq a$ and by the dual of Lemma A, $x(\bigvee b_\alpha) = \bigvee(xb_\alpha) \leq a$ whence $x \leq a : (\bigvee b_\alpha)$, and $a : (\bigvee b_\alpha)$ is the g. l. b. of the $a : b_\alpha$.

From now on we assume the associativity of the multiplication.

    ix) $(a : b) : c = a : cb$ and $(a :: b) :: c = a :: bc$.

    x) $(a : b) :: c = (a :: c) : b$.

    xi) $a(b : c) \leq ab : c$ and $(b :: c)a \leq ba :: c$.

    xii) $a : b \leq (a : c) : (b : c)$ and $a :: b \leq (a :: c) :: (b :: c)$.

The following examples are typical.

1. Let K be a groupoid with 0 element and let H be the groupoid of all subsets of K with 0 under complex multiplication. H is residuated: if $A, B \in H$, then $A : B = [x \in K \mid xB \subseteq A]$ and $A :: B = [y \in K \mid By \subseteq A]$.

2. Let R be an associative ring and S the semigroup of all ideals of R. S is residuated and $A : B$, $A :: B$ are the usual ideal quotients.

3. Let G be a group and S the lattice of all normal subgroups of G, the "product" of A, $B \in S$ being the commutator $[A, B]$. Then S is residuated.

2. Lattice-ordered semigroups

The reader is reminded that a *lattice-ordered groupoid*[2] has been defined as a groupoid H which is at the same time a lattice satisfying

(1) $$c(a \lor b) = ca \lor cb, \quad (a \lor b)c = ac \lor bc$$

for all $a, b, c \in H$. If H is (conditionally) complete and fulfils the infinite distributive laws

(2) $$a(\lor b_a) = \lor(ab_a), \quad (\lor b_a)a = \lor(b_a\, a),$$

it is said to be a *(conditionally) complete l. o. groupoid*. Note that now we need not postulate the monotony laws ($a \leq b$ implies $ca \leq cb$ etc.), for they follow at once from (1). Since we have not assumed the duals of (1) and they do not follow from (1), the duality principle fails for l. o. groupoids.

Some properties of l. o. groupoids may easily be inferred.

(a) $(a \land b)(a \lor b) \leq ab \lor ba$ and $(a \lor b)(a \land b) \leq ab \lor ba$, for $(a \land b)(a \lor b) = (a \land b)a \lor (a \land b)b \leq ba \lor ab$.

(b) If H has an identity e, then $a \lor b = e$ implies $a \land b = ab \lor ba$. For by (a) we have $\leq$, while $ba \leq ea = a$ etc. show that both ab and ba are smaller than or equal to a, b.

(c) If H has an identity e, then all $a_i \leq e$ satisfy[3]

$$a_1 a_2 \ldots a_n \leq a_1 \land a_2 \land \ldots \land a_n.$$

This follows by a trivial induction from $ab \leq a$, $ab \leq b$.

We now turn to considering the connections between l. o. and residuated groupoids.

(d) A residuated groupoid H which is at the same time a lattice is a l. o. groupoid. This follows immediately from

---

[2] In French mathematical literature l. o. semigroups are called *gerbier*; see e. g. DUBREIL-JACOTIN—LESIEUR—CROISOT [1].

[3] Brackets are omitted, because this is valid for arbitrary bracketing.

Lemma A in 1. This statement can obviously be generalized to complete lattices and complete l. o. groupoids.

(e) In a complete l. o. groupoid H, the residual $a : b$ exists if, and only if, some $x \in H$ satisfies $xb \leq a$. To prove the sufficiency of this condition, take the union u of all x_a satisfying $x_a b \leq a$. Then $ub = (\vee x_a)b = \vee(x_a b) \leq a$ and $u = a : b$. — In particular, a complete l. o. groupoid with 0 (which is the minimal element) or a l. o. groupoid with maximum condition and with 0 is residuated.

A number of results on l. o. groups can be carried over to l. o. semigroups under certain circumstances; for details we refer to DUBREIL-JACOTIN—LESIEUR—CROISOT [1].

3. The equivalence of Artin

In this section let S mean a residuated $\vee$-semigroup. We set up in S an equivalence relation A_t, called the *equivalence of* ARTIN, as follows.[4] Let $t \in S$ and define, for $a, b \in S$,

$$a \equiv b \; (A_t) \text{ if, and only if, } t : a = t : b.$$

The relation $_tA$ is similarly defined from left-residuals.

(A) A_t is an equivalence relation satisfying: $a \equiv b \; (A_t)$ implies $ca \equiv cb \; (A_t)$ and $a \vee c \equiv b \vee c \; (A_t)$ for all $c \in S$. These follow from ix) and Lemma C, respectively. If S is commutative, then A_t is a congruence relation on S.

(B) Every equivalence class mod A_t is convex and contains a maximal element. The maximal element equivalent to a is

$$a^\triangle = t :: (t : a).$$

For, clearly $a^\triangle \geq a$ and $a^\triangle \equiv a \; (A_t)$ on account of v), vi). If $a \equiv b \; (A_t)$, then

$$a^\triangle = t :: (t : a) = t :: (t : b) = b^\triangle \geq b.$$

Note that $a^\triangle = b^\triangle$ has the same meaning as $a \equiv b \; (A_t)$, because $a^\triangle = b^\triangle$ implies, in view of vi), $t : a = t : a^\triangle = t : b^\triangle = t : b$.

---

[4] This generalizes ARTIN's well-known equivalence relation introduced in commutative rings.

(C) The relation A_t can be characterized as an equivalence relation E_t with convex classes such that every class contains a left-residual of t which is the greatest element of the class. If this condition is fulfilled, then to $a \in S$ we can find one and only one $t :: x \in S$ with $a \equiv t :: x$ (E_t) and $a \leq t :: x$. Hence $a \leq t :: (t : a) \leq t :: [t : (t :: x)] = t :: x$ and so by convexity $t :: (t : a) = t :: x$. Consequently, $a \equiv t :: x$ (A_t) and E_t is finer than or equal to A_t. But each class mod A_t contains exactly one left-residual of t (namely, its maximal element), therefore the statement follows.

(D) The correspondence $a \to a^\triangle$ is a closure operation: $a \leq a^\triangle$; $a \leq b$ implies $a^\triangle \leq b^\triangle$; $(a^\triangle)^\triangle = a^\triangle$. Moreover, it satisfies

$$(a^\triangle \vee b^\triangle)^\triangle = (a \vee b)^\triangle.$$

The correspondence $a \to a^\triangledown = t : (t :: a)$ is likewise a closure operation. The $^\triangle$ closed elements and the $^\triangledown$ closed elements define a Galois correspondence

$$a^\triangle \to t : a^\triangle$$

with the inverse $a^\triangledown \to t :: a^\triangledown$ (see vi)).[5]

(E) $(t : a)a \equiv t$ (A_t) if, and only if, $(t : a) : (t : a) = t : t$. This follows immediately from ix).

If we specialize $t = e$, the identity of S, and at the same time assume that S is commutative, then $xa \equiv e$ (A_e) for all $a \in S$ with suitable x if, and only if, $a : a = e$ for all $a \in S$. To verify necessity, note that $e : xa = e : e = e$ implies $xa \leq e$, $x \leq e : a$. Now trivially $a : a \geq e$, while $e : (e : a) \geq a$ implies $e = e : xa = (e : a) : x \geq (e : a) : (e : a) = [e : (e : a)] : a \geq a : a$. Calling S *integrally closed* if

$$a : a = e \quad \text{for all } a \in S,$$

we arrive at the first part of

Theorem 1. (DUBREIL-JACOTIN [1].) *Let S be a commutative residuated $\vee$-semigroup with identity e. The factor*

---

[5] Interrelations between residuals and Galois correspondences have been discussed by· P. DUBREIL and R. CROISOT, *Collectanea Math.*, **7** (1954), 193—203.

semigroup S/A_e with respect to ARTIN'S *equivalence* A_e *is a (l. o.) group if, and only if, S is integrally closed. A_e is the only congruence relation* E *of S for which $S/$E is a group and e is the greatest element of its class.*

If E is a congruence relation of the stated kind, then to each $a \in S$ there is a $b \in S$ with $ab \equiv e$ (E) and $ab \leq e$. Thus $a \leq e : b$ and $ab \leq (e : b)b \leq e$. The classes mod E are, by preservation of unions, convex; thus $(e : b)b \equiv e$ (E). Hence $a \equiv e : b$ (E) and every class mod E contains a residual of e. Furthermore, $a \equiv b$ (E) implies $(e : b)a \equiv (e : b)b \equiv e$ (E), so $(e : b)a \leq e$ and $e : b \leq e : a$. By symmetry, $e : a = e : b$. Therefore E is finer than or equal to A_e. Because of the presence of residuals in the classes mod E, the stated equality is seen to hold. Q. E. D.

Observe that, in a V-semigroup S with identity e, $a : a$ is an idempotent element $\geq e$. In fact, $(a : a)a \leq a$ implies $(a : a)^2 a \leq a$, i. e., $(a : a)^2 \leq a : a$, while $ea = a$ implies $e \leq \leq a : a$ whence $a : a \leq (a : a)^2$. We conclude that S is necessarily integrally closed in the following cases: 1. S has no idempotent $\neq e$; 2. S is cancellative; 3. S is completely integrally closed.

In a residuated complete V-semigroup S the conditions : (i) S is integrally closed from the left; (ii) S is integrally closed from the right; (iii) S is completely integrally closed are equivalent. We show that (i) implies (iii). Let $a^n < b$ for $n = 1, 2, \ldots$; then $\mathsf{V}a^n = c$ exists and satisfies $ac \leq c$. Hence $a \leq c : c = e$.

If we also assume that in the V-semigroup S every non-void set of elements $\leq e$ contains a maximal member, then we can apply Theorem 3 in Chapter V to conclude that in this case every element of S may be uniquely written mod A_e as a product of powers (with positive or negative exponents) of "primes", i. e. elements p satisfying $p < e$ and $p \leq x \leq e$ implies $x \equiv p$ or $x \equiv e$ (A_e).

For further results on ARTIN'S equivalence relation we refer to DUBREIL [2] and MOLINARO [1], [2]. The latter author gave a systematic study of similar types of equivalence relation.

4. Elements with special properties

In this and in the following two sections we assume that S is a residuated l. o. semigroup and contains a *zero* 0 and a *universal element* u such that

$$0 \leq x \leq u, \quad x0 = 0x = 0, \quad xu \leq x, \quad ux \leq x \quad \text{for all } x \in S.$$

A typical example of such an S is the l. o. semigroup of all ideals of an associative ring. Our main purpose is to extend some important results in ideal theory to our present case. Here we must content ourselves with a discussion of some selected topics, for an extensive study falls outside the scope of the present treatment.[6]

As it is desirable to frame the argument with considerable generality, we shall base our discussion on the idea of operators. An *operator* Φ associates with each element x of S again an element $\Phi(x)$ of S such that

(1) $$x \leq \Phi(x) \quad \text{for all } x \in S.]$$

In certain cases we shall in addition assume that Φ satisfies

(2) $$x \leq \Phi(y) \text{ implies } \Phi(x) \leq \Phi(y) \text{ for all } x, y \in S$$

or

(3) $$\Phi(x \wedge y) = \Phi(x) \wedge \Phi(y) \quad \text{for all } x, y \in S.$$

An operator Φ satisfies (2) if, and only if, it is a *closure* operator:

$$\Phi(\Phi(x)) = \Phi(x) \text{ and } \Phi(x) \leq \Phi(y) \text{ if } x \leq y.$$

We shall call Φ *linear* if it satisfies (3).

By means of operators Φ we introduce the following concepts:[7]

A) An element $p \in S$ is Φ-*prime*, if

$$x_1 \ldots x_k \leq p \text{ implies } x_i \leq \Phi(p) \text{ for some } i.$$

---

[6] W. KRULL [*S. B. d. phys.-med. Soz. Erlangen*, **56** (1924), 47—63] was the first to discuss ideal-theoretic problems by lattice-theoretic methods.

[7] See the author's papers [4] and [8].

B) An element $q \in S$ is Φ-*primary*, if

$$x_1 \ldots x_k \leq q$$

implies that if $1 \leq i \leq k$, then

either $x_i \leq q$ or $x_j \leq \Phi(q)$ for some $j \neq i$.

Note that if this is true for $k = 2$, then it is true for all k. In fact, assume that for some $q \in S$ property B) holds with $k = 2$. If $x_1 \ldots x_k \leq q$ and $x_j \not\leq \Phi(q)$ for all $j \neq i$, then we get successively

$$x_2 \ldots x_k \leq q, \ldots, \ x_i \, x_{i+1} \ldots x_k \leq q, \ x_i \ldots x_{k-1} \leq q, \ldots,$$

and finally we arrive at $x_i \leq q$.

C) We call an $r \in S$ *right* Φ-*primal* if

$$x \leq \Phi(r) \text{ is equivalent to } r : x > r.$$

An element is Φ-*primal* if it is both right and left Φ-primal.[8]

If Φ is the identity operator, the symbol Φ will be omitted. [In this case we may take $k = 2$ in A).]

The connections between these notions are shown by

Lemma. *Every* Φ-*primal element is* Φ-*primary and every* Φ-*primary element is* Φ-*prime.*

If r is Φ-primal and $x_1 \ldots x_k \leq r$ where $x_j \not\leq \Phi(r)$ for $j \neq i$, then $r : x_j = r$ and $r :: x_j = r$ for $j \neq i$. Therefore $x_i \leq r$, r is Φ-primary. The second assertion is a trivial consequence of the inclusion $x \leq \Phi(x)$. It is not hard to illustrate by examples that none of the converse implications holds.

Note that *if r is right Φ-primal, then $\Phi(r)$ is a prime element.* Indeed, if $x_1 x_2 \leq \Phi(r)$, then $r : (x_1 x_2) > r$ which implies $(r : x_2) : x_1 > r$. Therefore $r : x_2 > r$ or $r : x_1 > r$ necessarily holds.

We also observe that if $x \in S$, $x \neq u$, has the property that $x = \Phi(x)$ (when x is called a Φ-*element*), then for x the properties of being Φ-prime, Φ-primary, Φ-primal and prime all coincide.

---

[8] Note that necessarily $r : x \geq r$ since the assumption that S contains a universal element implies that S is negatively ordered.

Illustrations for these concepts, together with the discussion of special cases, will be given in Section **6**. Now we consider the problem of finding conditions under which the intersection of a finite number of elements with a certain Φ-property has again the same Φ-property.

Theorem 2 (FUCHS [4].) *Let Φ be a closure operator. The intersection*

$$p = p_1 \wedge \ldots \wedge p_n$$

of a finite number of Φ-primes p_i is Φ-prime again if, and only if,

$$\Phi(p) = \Phi(p_j) \quad \text{for some } j.$$

If the intersection p of the p_i is Φ-prime, then[9] $p_1 \ldots p_n \leq\, \leq p_1 \wedge \ldots \wedge p_n = p$ implies $p_j \leq \Phi(p)$ for some j, and hence, by (2), $\Phi(p_j) \leq \Phi(p)$. But $p \leq p_j$ implies the converse inequality, thus $\Phi(p) = \Phi(p_j)$. Conversely, if $\Phi(p) = \Phi(p_j)$ for some j, then $x_1 \ldots x_k \leq p \leq p_j$ and the Φ-prime character of p_j implies $x_i \leq \Phi(p_j) = \Phi(p)$ for some index i.

Call the intersection $a = x_1 \wedge \ldots \wedge x_k$ *irredundant* if no x_i may be omitted, and *reduced*, if no x_i may be omitted or replaced by a greater element.

Theorem 3. (FUCHS [4].) *Let*

$$q = q_1 \wedge \ldots \wedge q_n$$

be an irredundant intersection of a finite number of Φ-primary elements q_j where Φ is a closure operator. A necessary and sufficient condition that q be Φ-primary is that

$$\Phi(q) = \Phi(q_j) \quad \text{for every } j.$$

Let q be Φ-primary. Clearly,

$$(q_1 \wedge \ldots \wedge q_{j-1} \wedge q_{j+1} \wedge \ldots \wedge q_n)q_j \leq q$$

for each j, and here the first factor is not $\leq q$ as irredundancy has been assumed. By hypothesis, $q_j \leq \Phi(q)$ whence $\Phi(q_j) \leq$

---

[9] We shall make use several times of the fact that a product is smaller than, or equal to, the intersection. This is a consequence of the presence of a universal element.

$\leq \Phi(q)$. Since $q \leq q_j$, the converse inclusion also holds. Conversely, let $\Phi(q) = \Phi(q_j)$ for every j, and $x_1 \ldots x_k \leq q$, $x_i \lneqq q$ for some i. Then $x_i \lneqq q_j$ for at least one index j, and therefore $x_1 \ldots x_k \leq q_j$ implies $x_l \leq \Phi(q_j) = \Phi(q)$ for some $l \neq i$. This shows q to be Φ-primary.

Theorem 4. (FUCHS [8].) *A reduced intersection*

$$r = r_1 \wedge \ldots \wedge r_n$$

of right Φ-primal elements r_i is likewise right Φ-primal if, and only if,

$$\Phi(r) = \Phi(r_j) \ \text{for some } j \ \text{and} \ \Phi(r) \geq \Phi(r_i) \ \text{for all } i.$$

Suppose that r is right Φ-primal. Since $r : a = (r_1 : a) \wedge \wedge \ldots \wedge (r_n : a)$ and the intersection is reduced, it follows that $a \leq \Phi(r)$ is equivalent to $a \leq \Phi(r_i)$ for some i. Hence $\Phi(r_i) \leq \Phi(r)$ for all i and $\Phi(r) \leq \Phi(r_j)$ for some j. To prove the converse, assume $\Phi(r)$ has the stated properties. Then $r : a > r$ is equivalent to $a \leq \Phi(r_j)$ for some j and thus to $a \leq \Phi(r)$. Q. E. D.

5. Unicity statements on meet decompositions

Our next aim is to study the decompositions of elements as intersections of a finite number of elements having the same Φ-property. The little that was assumed on the operators is naturally insufficient to prove the existence of such decompositions for every element. But the question as to whether such decompositions, if they exist at all, are in some sense unique can be answered in full generality. The results run parallel to those known in commutative ideal theory.

Call a meet decomposition $a = x_1 \wedge \ldots \wedge x_n$ of the element a in S (with the same hypothesis on S as in **4**) with elements x_i having some Φ-property *short* if it is irredundant and no subset of the x_l has a meet which has again the same Φ-property. From an arbitrary meet decomposition we clearly can get a short one by omitting some components and suitably combining others.

Theorem 5. (FUCHS [4].) *Let Φ be a linear closure operator. If*

$$a = p_1 \wedge \ldots \wedge p_n = p_1^* \wedge \ldots \wedge p_m^*$$

are two short decompositions of $a \in S$ into the meet of Φ-primes p_i, p_j^*, then $n = m$ and, after possibly re-arranging the terms, we have

$$\Phi(p_i) = \Phi(p_i^*) \quad \text{for all } i.$$

For every j, the inclusion $p_1 \ldots p_n \leq a \leq p_j^*$ implies $p_i \leq \Phi(p_j^*)$ for some $i = i(j)$. Hence $\Phi(p_i) \leq \Phi(p_j^*)$. The same reasoning applied to p_i instead of p_j^* shows that $\Phi(p_k^*) \leq \Phi(p_i)$ for some $k = k(i)$. Since $\Phi(p_j^* \wedge p_k^*) = \Phi(p_j^*) \wedge \Phi(p_k^*) = \Phi(p_k^*)$, by virtue of Theorem 2 the intersection $p_j^* \wedge p_k^*$ is again Φ-prime, and so only $j = k$ is possible. Thus $\Phi(p_i) = \Phi(p_j^*)$ and the proof is completed.

Theorem 6. (FUCHS [4].) *Assume that Φ is a linear closure operator. If an element a of S has two short meet decompositions*

$$a = q_1 \wedge \ldots \wedge q_n = q_1^* \wedge \ldots \wedge q_m^*$$

with Φ-primary components q_i and q_j^*, then $n = m$ and the elements $\Phi(q_1), \ldots, \Phi(q_n)$ are, up to order, equal to the elements $\Phi(q_1^*), \ldots, \Phi(q_m^*)$.

If $\Phi(q_i)$ is a minimal one amongst $\Phi(q_1), \ldots, \Phi(q_n)$, and say $\Phi(q_i) = \Phi(q_i^*)$, then $q_i = q_i^*$.

In view of Theorem 3 and the assumption on Φ, the elements $\Phi(q_1), \ldots, \Phi(q_n)$ are different, and so are $\Phi(q_1^*), \ldots, \Phi(q_m^*)$. Select a maximal one, say $\Phi(q_1)$, amongst all of $\Phi(q_i)$, $\Phi(q_j^*)$, and form the residual $a : q_1 = q_1 : q_1 \wedge \ldots \wedge q_n : q_1$. Then $q_1 \nleq \Phi(q_i)$ for $i > 1$ and so from $(q_i : q_1)q_1 \leq q_i$ we infer $q_i : q_1 \leq q_i$, i. e. $q_i : q_1 = q_i$ for $i > 1$. Thus $a : q_1 = q_2 \wedge \ldots \wedge q_n > a$ by irredundancy, and therefore $a : q_1 = q_1^* : q_1 \wedge \ldots \wedge q_m^* : q_1$ shows that $q_j^* : q_1 > q_j^*$ for some j. For this j we obtain from $(q_j^* : q_1)q_1 \leq q_j^*$ that $q_1 \leq \Phi(q_j^*)$ whence $\Phi(q_1) \leq \Phi(q_j^*)$. This proves that the maximal ones among $\Phi(q_i)$ and those among $\Phi(q_j^*)$ are the same.

Let now $\Phi(q_1) = \Phi(q_1^*)$ be a maximal one amongst the $\Phi(q_i)$, $\Phi(q_j^*)$. Then $q_1 \wedge q_1^*$ is again Φ-primary with $\Phi(q_1 \wedge q_1^*) =$

$= \Phi(q_1) = \Phi(q_1^*)$. Therefore $q_i : (q_1 \wedge q_1^*) = q_i$ for $i > 1$ and $q_j^* : (q_1 \wedge q_1^*) = q_j^*$ for $j > 1$. Consequently,

$$a_1 = a : (q_1 \wedge q_1^*) = q_2 \wedge \ldots \wedge q_n = q_2^* \wedge \ldots \wedge q_m^*,$$

and an easy induction establishes the first assertion of the theorem.

Finally suppose that $\Phi(q_n) = \Phi(q_n^*)$ is minimal amongst the $\Phi(q)$'s. Define $c = q_1 \wedge \ldots \wedge q_{n-1} \wedge q_1^* \wedge \ldots \wedge q_{n-1}^*$ and form $a : c$. Clearly, $q_i : c = q_i^* : c = u$ for all $i < n$. The inequality

$$(q_n : c)q_1 \ldots q_{n-1} q_1^* \ldots q_{n-1}^* \leq (q_n : c)c \leq q_n$$

together with $q_i \lessgtr \Phi(q_n)$, $q_i^* \lessgtr \Phi(q_n)$ $(i < n)$ implies $q_n : c \leq$ $\leq q_n$. That is $q_n : c = q_n$ and, similarly, $q_n^* : c = q_n^*$. Hence it follows that $q_n = q_n : c = a : c = q_n^* : c = q_n^*$, as claimed.

Theorem 7. (FUCHS [8].) *Let $a \in S$ have two reduced meet decompositions into a finite number of right Φ-primal elements,*

$$a = r_1 \wedge \ldots \wedge r_n = r_1^* \wedge \ldots \wedge r_m^*.$$

Then the (different) maximal ones amongst $\Phi(r_1), \ldots, \Phi(r_n)$ and those amongst $\Phi(r_1), \ldots, \Phi(r_m^*)$ are the same.

The relation $r_i : \Phi(r_i) > r_i$ and reducedness imply

$$a : \Phi(r_i) = r_1 : \Phi(r_i) \wedge \ldots \wedge r_n : \Phi(r_i) > a.$$

Therefore $r_j^* : \Phi(r_i) > r_j^*$ for at least one index j, that is, $\Phi(r_i) \leq \Phi(r_j^*)$. Starting with this j, we obtain $\Phi(r_j^*) \leq \Phi(r_k)$ for some k, $1 \leq k \leq n$. If $\Phi(r_i)$ is maximal, then necessarily $\Phi(r_i) = \Phi(r_j^*) = \Phi(r_k)$, and $\Phi(r_j^*)$ must again be maximal. Q. E. D.

A stronger statement can be proved if we assume that Φ satisfies the following condition: if $r_1, \ldots, r_n$ are right Φ-primal elements with $\Phi(r_1) \geq \Phi(r_j)$ for all j, and $r_1 \wedge \ldots \wedge r_n = r$ is a reduced intersection, then $\Phi(r) = \Phi(r_1)$.[10] In fact, it follows from Theorem 4 that in this case the $\Phi(r_i)$ that belong to a reduced and short meet decomposition are all maximal and distinct. Therefore the set $\Phi(r_1), \ldots, \Phi(r_n)$ is uniquely determined by a.

[10] The operator A which will be considered in **6** has this property.

To sum up, we may say, broadly speaking, that many unicity statements in ideal theory retain their validity in the much more general case here discussed.

6. Meet decompositions of elements

Our next task is to seek conditions that must be imposed on the operator Φ in order to guarantee the existence of meet decompositions of all elements of the l. o. semigroup S into Φ-prime, Φ-primary or Φ-primal components. It turns out that some additional hypothesis on S is necessary. But to make our treatment easier, we shall not aim at the greatest possible generality; instead we shall content ourselves with the case when some finiteness condition is assumed. The most important one is the *maximum condition*: every non-void subset of S contains a maximal element.

Throughout this section S is a l. o. semigroup with 0 and universal element u $[0 \leq x \leq u, 0x = x0 = 0, xu \leq x, ux \leq \leq x$ for all $x \in S]$ in which the maximum condition holds. Note that the existence of residuals in S is a simple consequence of the presence of 0 and of the maximum condition.

Call an element a of S *meet-irreducible* if

$$a = b \wedge c \ (b, c \in S) \text{ implies } b = a \text{ or } c = a.$$

Lemma. *Every element of S is the intersection of a finite number of meet-irreducible elements. The same holds for reduced intersections.*

Assume that the set of elements of S for which this property fails is not empty and pick out a maximal element a from this set. Since for irreducible elements the assertion of the lemma holds trivially, we have $a = b \wedge c$ with $b > a$ and $c > a$. But then both b and c are intersections of a finite number of meet-irreducible elements and so the same must be true for a, contrary to hypothesis. Since it is easily verified that $a = b \wedge c$ may be made reduced and it remains reduced if for b and c reduced intersections are substituted, the same inference is applicable to prove the second assertion.

Proposition 8. *Let S be a l. o. semigroup of the stated kind. Every element of S can be written as an intersection of elements having some fixed property if, and only if, every irreducible element of S has this property.*

The non-trivial part follows from the preceding lemma.

Now we specialize the operators Φ and inquire what kind of meet decompositions can be established.

I. Let $\Phi = \mathsf{I}$ be the identity operator, $\mathsf{I}(x) = x$ for all $x \in S$. In this case, for elements $x \neq u$, the properties of being prime, I-prime, I-primary, and I-primal, respectively, are identical.[11] The meet-irreducible elements are in general not prime, and therefore decompositions of the desired kind for all elements of S cannot be established.

II. Let $\Phi = \mathsf{P}$ be the *radical* operator: $\mathsf{P}(x)$ is the join of all $a \in S$ such that

$$a^k \leq x \quad \text{for some natural integer } k.$$

Since, owing to the maximum condition, $\mathsf{P}(x)$ is the join of a finite number of the a's, $\mathsf{P}(x) = a_1 \vee \ldots \vee a_n$ with $a_i^{k_i} \leq x$, we have

$$\mathsf{P}(x)^k = \vee\, a_{i_1} \ldots a_{i_k} \leq x$$

whenever $k \geq k_1 + \ldots + k_n$. Hence $\mathsf{P}(x)$ can alternatively be defined as the unique greatest element a which has a power $\leq x$. $\mathsf{P}(x)$ is semi-prime.[12]

It is readily seen that P is a closure operator. Evidently, $\mathsf{P}(x \wedge y) \leq \mathsf{P}(x) \wedge \mathsf{P}(y)$. If $a^r \leq x$ and $a^s \leq y$, then $a^t \leq \leq x \wedge y$ with $t = \max(r, s)$, so that $\mathsf{P}(x) \wedge \mathsf{P}(y) \leq \mathsf{P}(x \wedge y)$ and P is linear:

$$\mathsf{P}(x \wedge y) = \mathsf{P}(x) \wedge \mathsf{P}(y).$$

Call a prime element p a *minimal prime belonging to* x if $x \leq p$ but no prime p' satisfies $x \leq p' < p$.

Proposition 9. *To every element $x \in S$ there belong only a finite number of minimal primes, and their intersection is $\mathsf{P}(x)$.*

---

[11] Cf. a remark after the lemma in **4**.

[12] An element t is called *semi-prime* if $a^n \leq t$ for some natural integer n implies $a \leq t$.

If a prime p satisfies $x \leq p$, then it also satisfies $P(x) \leq p$, for $P(x)^k \leq x \leq p$. If $y \leq P(x)$, then no power of y is $\leq x$, and therefore there exists an element $p \in S$ maximal with respect to the properties $x \leq p$ and $y^k \leq p$ for no k. This p must be prime, for otherwise $ab \leq p$, $a \leq p$, $b \leq p$ imply $y^{k_1} \leq a \vee p$, $y^{k_2} \leq b \vee p$ for some k_i, and so we get $y^{k_1+k_2} \leq$ $\leq (a \vee p)(b \vee p) \leq p$. Therefore

$$P(x) = \bigwedge_{x \leq p_\nu} p_\nu \quad \text{(with primes } p_\nu\text{)}.$$

We show that here a finite number of primes suffice to represent $P(x)$. If $P(x)$ is irreducible and $ab \leq P(x)$, then $ab \leq p_\nu$ for every p_ν, thus $a \leq p_\nu$ or $b \leq p_\nu$. Writing

$$a' = \bigwedge_{a \leq p_\nu} p_\nu \quad \text{and} \quad b' = \bigwedge_{b \leq p_\nu} p_\nu,$$

we get $P(x) = a' \wedge b'$ where by irreducibility either $a' = P(x)$ or $b' = P(x)$, that is to say, either $a \leq P(x)$ or $b \leq P(x)$, and so $P(x)$ is prime. If $P(x)$ is reducible and $P(x) = a \wedge b$ is a reduced decomposition with $P(x) < a$, $P(x) < b$, then $ab \leq P(x)$ and as before we get $a \leq a'$ and $b \leq b'$ with $P(x) = a' \wedge b'$. By reducedness we obtain $a = a'$ and $b = b'$, and we conclude that both a and b are intersections of primes. By the maximum condition we can use an induction to obtain a reduced representation

$$P(x) = p_1 \wedge \ldots \wedge p_r \quad \text{(with primes } p_i\text{)}.$$

For each prime p with $x \leq p$ we have $(p_1 \ldots p_r)^k \leq P(x)^k \leq$ $\leq x \leq p$ whence $p_i \leq p$ for some p_i, that is, just the $p_1, \ldots, p_r$ are the minimal primes belonging to x. Q. E. D.

$p \in S$ is P-*prime if, and only if,* $P(p)$ *is prime* (or, in other words, only one minimal prime belongs to p). For if p is P-prime and $ab \leq P(p)$, then $(ab)^k \leq p$ for a certain k. Hence either $a \leq P(p)$ or $b \leq P(p)$. Conversely, if $P(p)$ is prime and $a_1 \ldots a_k \leq p$, then $a_1 \ldots a_k \leq P(p)$. This implies $a_i \leq P(p)$ for some i.

The notions P-*primary and* P-*primal are identical.*[13] It suffices to show that a P-primary element q ($\neq u$) is necessarily

---

[13] We may consider u here as a primal element.

P-primal. If $a \leq \mathsf{P}(q)$, then $a^k \leq q$ and if here k is as small as possible, then $a^{k-1} \not\leq q$ and $q : a > q$. On the other hand, if $q : a > q$, then from $(q : a)a \leq q$ and the P-primary character of q we get $a \leq \mathsf{P}(q)$.

Meet-irreducibility implies the P-primary property under additional assumptions.

Proposition 10. (WARD—DILWORTH [2].) *If S is modular as a lattice and if $a, b \in S$ guarantee the existence of integers k, l such that*

$$a^k \wedge b \leq ab \quad and \quad a \wedge b^l \leq ab,$$

then every meet-irreducible element of S is P-primary.

Let $ab \leq q$, $a \not\leq q$ where q is meet-irreducible. By hypothesis $q : b \wedge b^l \leq (q : b)b \leq q$ for some l, whence

$$q = q \vee [q : b \wedge b^l] = q : b \wedge (q \vee b^l).$$

Now $q : b > q$ implies $q \vee b^l = q$, i. e. $b^l \leq q$, $b \leq \mathsf{P}(q)$. Similarly, $ab \leq q$, $b \not\leq q$ imply $a \leq \mathsf{P}(q)$.

III. Define $\mathsf{P}_2(x)$ as the join of all $a \in S$ which satisfy: there exist natural integers $k_0, k_1, \ldots, k_n$ $(n \geq 0)$ and elements $v_1, \ldots, v_n$ of S such that

$$a^{k_0} v_1 a^{k_1} v_2 \ldots v_n a^{k_n} \leq x \quad \text{and} \quad x : v_i = x \quad (i = 1, 2, \ldots, n).$$

Obviously, every v_i can be replaced by their product and every k_i by the maximal k_i. As $y \leq x$ and $yv \leq x$ are equivalent whenever $x : v = x$, the definition of $\mathsf{P}_2(x)$ can be formulated more simply as the join of all $a \in S$ satisfying

$$(a^k v)^n \leq x \quad \text{for integers } k, n > 0 \text{ and } x : v = x.$$

If both a_1 and a_2 possess this property, then so does $a_1 \vee a_2$. For if $(a_1^k v_1)^m \leq x$ and $(a_2^l v_2)^n \leq x$ with $x : v_i = x$, and if $m \geq n$, then $[(a_1 \vee a_2)^{k+l} v_1 v_2]^m \leq x$ with $x : v_1 v_2 = x$. Hence $\mathsf{P}_2(x)$ is the (unique) maximal a having the stated property. $\mathsf{P}_2(x)$ is called the *secondary radical*[14] of x. Manifestly,

$$\mathsf{P}(x) \leq \mathsf{P}_2(x) \quad \text{for all } x \in S,$$

and in the commutative case equality holds.

---

[14] The secondary and tertiary radicals have been discovered by LESIEUR and CROISOT [1].

A similar argument as above shows that $p \in S$ is P_2-prime if, and only if, $P_2(p)$ is prime,[15] and that P_2-primary elements are necessarily right P_2-primal. It is not hard to give a representation of $P_2(x)$ as the intersection of certain primes p with $x \leq p$.

From the definition it is immediately seen that if $x = = x_1 \wedge \ldots \wedge x_n$ is a reduced intersection, then $P_2(x) \leq \leq P_2(x_1) \wedge \ldots \wedge P_2(x_n)$. In view of Theorem 4, a reduced intersection $r = r_1 \wedge \ldots \wedge r_n$ of right P_2-primal elements r_i can be right P_2-primal only if $P_2(r_1) = \ldots = P_2(r_n)$. But if this condition is satisfied then $P_2(r) = P_2(r_i)$ (note that $r_i : v = = r_i$ implies $r_j : v = r_j$), so that with $\Phi = P_2$, the condition of Theorem 4 is $P_2(r_1) = \ldots = P_2(r_n)$. The conclusion of Theorem 7 can be formulated for short and reduced decompositions as follows: the $P_2(r_1), \ldots, P_2(r_n)$ are different and the maximal ones among them are uniquely determined by a.

IV. Next let $P_3(x)$ be the union of all $a \in S$ such that

$$x : a \wedge b \leq x \quad \text{implies} \quad b \leq x.$$

It is readily seen that together with a_1 and a_2 also $a_1 \vee a_2$ has this property. Thus $P_3(x)$ is the (unique) maximal a with the stated property; it is called the *tertiary radical* of x.

Proposition 11. (LESIEUR—CROISOT [1].) *If S is modular as a lattice, then every meet-irreducible element is right P_3-primal.*

If x is meet-irreducible and $x : a \wedge b \leq x$, then $x = = (x : a \wedge b) \vee x = x : a \wedge (b \vee x)$. Hence $x : a > x$ implies $b \vee x = x$, $b \leq x$, that is, $a \leq P_3(x)$. The converse implication is obvious.

Let now $x = x_1 \vee \ldots \vee x_n$ and $a = P_3(x_1) \vee \ldots \vee P_3(x_n)$. If $x : a \wedge b \leq x$, then

$$(x_1 : a) \wedge (x_2 \wedge \ldots \wedge x_n) : a \wedge b \leq x_1$$

implies $(x_2 \wedge \ldots \wedge x_n) : a \wedge b \leq x_1$ and $(x_2 \wedge \ldots \wedge x_n) : a \wedge b \leq x_1 : a$. Therefore

$$(x_2 \wedge \ldots \wedge x_n) : a \wedge b = x : a \wedge b \leq x.$$

---

[15] Use the fact that $x : v = x$ implies that $v \leq P_2(x)$ is impossible, for otherwise also $x : P_2(x) = x$, which leads to an obvious contradiction.

A repeated application of this argument leads to $b \leq x$, proving that $a \leq \mathsf{P}_3(x)$. Now if we assume that $r = r_1 \wedge \ldots \wedge r_n$ is a reduced intersection of right P_3-primal elements r_i such that $a = \mathsf{P}_3(r_1) = \ldots = \mathsf{P}_3(r_n)$, then $r : c > r$ is equivalent to $c \leq a$, whence necessarily $\mathsf{P}_3(r) \leq a$. Combining these observations with Theorem 4 it follows that the r now considered is again right P_3-primal with $\mathsf{P}_3(r) = a$. The conclusion of Theorem 7 can be formulated as when $\Phi = \mathsf{P}_2$.

V. Let $\mathsf{A}(x)$ be the join of all $a \in S$ such that[16]

$$x : b > x \text{ (for some } b \in S) \text{ implies } x : (a \vee b) > x.$$

If a_1 and a_2 have this property, then $a_1 \vee a_2$ has it too, thus $\mathsf{A}(x)$ can be characterized as the greatest a with the stated property. Another useful characterization is given by the next

Proposition 12. (FUCHS [8].) *For a given $x \in S$, let p run over all elements of S for which*

(i) $x : p > x$,
(ii) $p' > p$ *implies* $x : p' = x$.

These p's are primes, are finite in number and their intersection is $\mathsf{A}(x)$.

Assume that $ab \leq p$. Then clearly $(a \vee p)(b \vee p) \leq p$ and $x < x : p \leq [x : (b \vee p)] : (a \vee p)$. Hence not both of $x : (b \vee p) = x$ and $x : (a \vee p) = x$ can be valid, that is, by (ii), either $b \leq \leq p$ or $a \leq p$, and so p is prime.

The relation $x : p > x$ implies $x : (\mathsf{A}(x) \vee p) > x$ whence, by (ii) we get $\mathsf{A}(x) \leq p$. Let $y \in S$ satisfy $y \leq p$ for all p satisfying (i) and (ii). Because of the maximum condition, to every b with $x : b > x$ there exists a p with $b \leq p$ that satisfies (i) and (ii). Then $y \vee b \leq p$ and hence $x : (y \vee b) \geq x : p > x$, showing that $y \leq \mathsf{A}(x)$. Thus $\mathsf{A}(x)$ is the intersection of all p's satisfying (i) and (ii). As in the proof of Proposition 9 it follows that

$$\mathsf{A}(x) = p_1 \wedge \cdots \wedge p_r$$

for a finite number of primes $p_1, \ldots, p_r$ satisfying (i) and (ii).

---

[16] According to this definition the operator A is not left-right symmetric.

If p is an element with (i) and (ii), then $p_1 \wedge \ldots \wedge p_r \leq p$ implies $p_1 \ldots p_r \leq p$ whence $p_i \leq p$ for some i and $p_i = p$ because of (ii). This shows that there exists no p other than $p_1, \ldots, p_r$. The proof is completed.

From Propositions 9 and 12 we see that

$$P(x) \leq A(x) \quad \text{for all } x \in S.$$

The general definition of right Φ-primality shows that a right A-primal element $r \in S$ is one satisfying:

$$r : a > r \text{ and } r : b > r \text{ imply } r : (a \vee b) > r,$$

or, in other words, it cannot be represented as the intersection of two of its right residuals unless one of them equals to it. The following conclusion is now obvious:

Proposition 13. *A meet-irreducible element is right* A-*primal.*

It follows readily that x is a right A-primal element if, and only if, only one p exists (defined by Proposition 12), or equivalently, $A(x)$ is prime.

An A-*primary element is right* A-*primal.* If q is A-primary and $q : a > q$, then in view of $(q : a)a \leq q$ we get $a \leq A(q)$. Hence $x : A(x) > x$ implies the assertion.

Next assume that $x = x_1 \wedge \ldots \wedge x_n$ is a reduced intersection. Then $x : a > x$ if, and only if, $x_i : a > x_i$ for some index i. Thus the primes satisfying (i) and (ii) of Proposition 12 relative to x are just the maximal ones among the primes satisfying the same conditions relative to some x_i. We conclude that A is neither a closure operator nor a linear operator. It also follows that a reduced intersection

$$r = r_1 \wedge \ldots \wedge r_n$$

of right A-primal elements r_i is itself right A-primal if, and only if, there is only one maximal prime among the primes $A(r_1), \ldots, A(r_n)$ (cf. Theorem 4). Hence one gets from the lemma and Proposition 13 the existence of reduced and short decompositions of the elements a of S into right A-primal elements with incomparable primes $A(r_i)$. By virtue of Theorem 7, the primes $A(r_i)$ are uniquely determined by a.

For further investigations of l. o. semigroups and groupoids we refer to the following papers: DUBREIL-JACOTIN [3], FUCHS [8], KERSTAN [1], KRISHNAN [2], LESIEUR [3], [4], MURATA [2], [3]. Several results in additive ideal theory can be carried over to this general case. A further generalization, which proceeds in the module-theoretic direction, has been given by LESIEUR—CROISOT [1].

STEINFELD [1] considers somewhat more general systems than l. o. semigroups; these are supposed to have the same properties as the system of all subrings of an associative ring.

The concept of nilpotency can also be discussed in l. o. semigroups; cf. BIRKHOFF [3]. STELLECKIĬ [1] investigates nilpotency in lattices in which multiplication is defined (but they are not necessarily l. o. groupoids, since distributivity is not assumed). He defines (right) normal and central systems etc. and shows that several results of the theory of groups and Lie rings may be carried over to this general case.

PROBLEMS

Here we list some unsolved problems concerning partially ordered groups, rings and semigroups.

1. Find necessary and sufficient conditions on a group G in order that there exist a directed (or an isolated directed) order on G.

2. What groups G have the property that every partial order of G can be extended to a directed order of G?

3. In which groups can every partial order be extended to an isolated partial order?

4. (BIRKHOFF [3].) When is an abstract group isomorphic to a l. o. group?

5. Characterize (group-theoretically) the groups in which every partial order can be strengthened to a lattice-order.

6. Give necessary and sufficient conditions on a directed set that it be o-isomorphic to (a) a p. o. group; (b) a p. o. ring; (c) a p. o. semigroup.

7. When is a (complete) distributive lattice lattice-isomorphic to a (complete) l. o. group? And to the positive cone of a l. o. group?

8. Let G be a directed (l. o.) group such that every f. o. subgroup is commutative. When can we conclude that G is commutative as well?

9. (a) Find necessary and sufficient conditions for the existence of a full order on an O-group (O^*-group) G such that a prescribed subgroup H of G should be convex. [If H is normal, then G/H must be an O-group.]

(b) The analogous problem for rings.

10. Let Σ be a system of subgroups in a group G. Under what conditions is it possible to define a directed order (or lattice-order) on G such that the subgroups in Σ should be convex? Or that Σ should be the set of all convex subgroups of G?

11. Call a subgroup C of an O-group (O^*-group) G *absolutely convex* if it is convex in every full order of G. Discuss the properties of absolutely convex subgroups.[1]

12. (a) Describe the f. o. groups B which have the property that if B is a normal convex subgroup in some f. o. group G, then G is the lexicographic product of a f. o. group A and B.[2]

(b) Find f. o. groups A for which every f. o. group G whose o-epimorphic image is A splits into the lexicographic product of A and the kernel of the o-epimorphism.

(c) The same questions for Abelian groups only.

13. (G. Grätzer.) Use the following general notion instead of m-direct product and lexicographic σ-product to obtain a suitable common generalization of the refinement Theorems 8 and 9 in Chapter I.

Let Λ be a f. o. index set such that

(i) Λ is decomposed into pair-wise disjoint subsets Λ_μ ($u \in \mathsf{M}$);

(ii) there is given an ideal $\mathcal{I}$ of the Boolean algebra of all subsets of M such that all finite subsets of M belong to $\mathcal{I}$;

(iii) for each   , there is given an ideal $\mathcal{I}_\mu$ of the Boolean algebra of all subsets of Λ_μ such that $\mathcal{I}_\mu$ contains only well-ordered subsets and includes all finite subsets of Λ_μ.

In the complete direct product

$$G = \underset{\lambda \in \Lambda}{\coprod}^* G_\lambda$$

of the directed (l. o.) groups G_λ we define the subgroup $H = H(\mathcal{I}, \mathcal{I}_\mu)$ (the "mixed direct—lexicographic" product) as the set of all vectors $\langle \ldots, g_\lambda, \ldots \rangle$ in G ($g_\lambda \in G_\lambda$) satisfying

$$\Lambda(g) \cap \Lambda_\mu \in \mathcal{I}_\mu \text{ for every } u; \quad [\iota \in \mathsf{M} \mid \Lambda(g) \cap \Lambda_\mu \neq \varnothing] \in \mathcal{I}$$

----

[1] In the group G generated by two elements a, b with the defining relations $a^{-1}ba = b^2$, the normal subgroup generated by b is absolutely convex. Another example is the group of all 2×2 matrices of the form $\begin{pmatrix} a & b \\ 0 & 1 \end{pmatrix}$ with a, b in a f. o. field and $a > 0$. The matrices $\begin{pmatrix} 1 & b \\ 0 & 1 \end{pmatrix}$ form a subgroup which is absolutely convex. (B. H. Neumann.)

[2] The second example in footnote [1] seems to be such a group.

where

$$\Lambda(g) = [\lambda \in \Lambda \mid g_\lambda \neq e],$$

and we put $g \geq e$ for some $g \in H$ if, and only if, for every μ, the first component g_λ distinct from e with $\lambda \in \mathscr{S}_\mu$ is greater than e.

14. (a) Try to derive VINOGRADOV's theorem that the free product of O-groups is likewise an O-group from the conditions of Chapter III, **2**.

(b) For what kinds of amalgamated free products can a similar result be established?

15. Is the direct product of O^*-groups again an O^*-group?

16. (B. H. NEUMANN.) Let G be an O-group and $a \in G$. When can G be embedded in an O-group H containing x satisfying $x^2 = a$? In which cases can a prescribed full order of G be extended to G?

17. (B. H. NEUMANN.) Let us call a group U *universal* for a class $\mathscr{C}$ of groups if U contains to every $G \in \mathscr{C}$ an isomorphic copy. Does there exist a countable group that is universal for the class of countable O-groups? [There is no countable f. o. group universal for the class of countable f. o. (Abelian) groups.]

18. (a) (B. H. NEUMANN.) If the number of full orders of an O-group is finite, must it be a power of 2? And if it is infinite, must it be a power of 2?

(b) The same questions for O-rings.[3]

(c) What are the O-groups (O-rings) with only a finite number of full orders?

19. Describe the group-theoretical structure of finitely generated O^*-groups. Investigate group-theoretical properties of O^*-groups in general. Does there exist an algebraically simple O^*-group?

20. Characterize the O^*-groups in which every subgroup is again an O^*-group. In general, which subgroups of O^*-groups are again O^*-groups?

---

[3] Cf. KLINGENBERG [1].

21. Assume that a group G has the property that every i s o l a t e d partial order of G can be extended to a full order of G. Need G then be an O^*-group?

22. Extend the results of Chapter III to loops (quasi-groups).

23. (BIRKHOFF [3].) Characterize abstractly the free l. o. group with two (and more generally, with a finite number of) generators.

24. Define partial cyclical order by properly modifying the axioms of cyclical order in Chapter IV, **6**, and develop a theory of partially cyclically ordered groups.

25. Establish ring-theoretical properties of O-rings and O^*-rings.

26. Enumerate the characterizing properties of the system of all convex subrings (left ideals, ideals) of a f. o. ring.

27. (G. GRÄTZER.) Let R be an (associative) O-ring. How can one characterize the set of all elements of R that are totally positive in the sense that they are positive in every full order of R?

28. Develop a theory for alternative, Jordan, and more generally, power-associative O-rings (in particular, division O-rings).

29. Does there exist a polynomial identity which implies that an (associative) O-ring must be an O^*-ring? [In the case of groups, $xy = yx$ is such an identity.]

30. (a) What can be saved of the theory of real closed commutative fields for fields which are maximal O-fields in the sense that there exist no O-fields of finite degree > 1 over them and which admit only a single full order?

(b) The analogous question for commutative and associative rings.

31. Describe the directed orders of the fields of complex numbers and quaternions.

32. (a) Define the O-radical of an abstract ring R as the intersection of all ideals I such that R/I is an O-ring. Try to determine the O-radical in terms of usual ring-theoretical concepts.

(b) What is the O-radical of a l. o. ring R if I are supposed to be L-ideals and R/I f. o. under the induced order? [This might measure the deviation from being an F-ring.]

(c) The corresponding problems concerning groups.

33. Discuss the different types of radical in non-associative f. o. (l. o.) rings.

34. When is an abstract ring isomorphic to a l. o. ring?

35. Characterize as abstract groups the multiplicative groups of O-fields.

36. Extend the results of Chapter VII to semirings.

37. Discuss l. o. semirings.

38. Develop a theory for l. o. modules over l. o. rings.

39. Establish conditions on a system of linear inequalities in f. o. or l. o. rings (fields) in order that it should possess a solution in the ring (or some f. o. extension of the ring).

40. Investigate bilinear forms in f. o. (l. o.) associative rings.

BIBLIOGRAPHY

ALBERT, A. A. [1] On ordered algebras, *Bull. Amer. Math. Soc.*, **46** (1940), 521—522. — [2] A property of ordered rings, *Proc. Amer. Math. Soc.*, **8** (1957), 128—129.

(ALIMOV) А л и м о в, Н. Г. [1] Об упорядоченных полугруппах (On ordered semigroups), Изв. Акад. Наук СССР, **14** (1950), 569—576.

ALLING, N. L. [1] On ordered divisible groups, *Trans. Amer. Math. Soc.*, **94** (1960), 498—514.

ARTIN, E. [1] Über die Zerlegung definiter Funktionen in Quadrate, *Abh. Math. Sem. Hamb. Univ.*, **5** (1926), 100—115.

ARTIN, E.—SCHREIER, O. [1] Algebraische Konstruktion reeller Körper, *Abh. Math. Sem. Hamb. Univ.*, **5** (1926), 85—99.

BAER, R. [1] Über nicht-archimedisch geordnete Körper, *S.-B. Heidelberger Akad. Wiss.*, **1927**, 8. Abh., 3—13. — [2] Zur Topologie der Gruppen, *Journ. reine u. angew. Math.*, **160** (1929), 208—226.

BANASCHEWSKI, B. [1] Totalgeordnete Moduln, *Archiv Math.*, **7** (1956), 430—440. — [2] Über die Vervollständigung geordneter Gruppen, *Math. Nachrichten*, **16** (1957), 51—71.

BAUER, H. [1] Geordnete Gruppen mit Zerlegungseigenschaft, *S.-B. Bayer. Akad. Wiss. Math.-Nat. Klasse*, **1958**, 25—36.

BIRKHOFF, G. [1] Lattice-ordered groups, *Annals Math.*, **43** (1942), 298—331. — [2] Lattice-ordered Lie groups, *Speiser Festschrift* (Zürich, 1945), 209—217. — [3] *Lattice theory*, 2nd ed. (New York, 1948). — [4] Groupes réticulés, *Ann. Inst. Henri Poincaré*, **11** (1949), 241—250.

BIRKHOFF, G.—PIERCE, R. S. [1] Lattice-ordered rings, *Anais Acad. Brasil. Ci.*, **28** (1956), 41—69.

BOURBAKI, N. [1] *Algèbre*, Chap. VI. Groupes et corps ordonnés (Paris, 1952).

BRAINERD, B. [1] On a class of lattice-ordered rings, *Proc. Amer. Math. Soc.*, **8** (1957), 673—683.

BRITTON, J. L.—SHEPPERD, J. A. H. [1] Almost ordered groups, *Proc. London Math. Soc.*, **1** (1951), 188—199.

BRUCK, R. H. [1] *A survey of binary systems* (Berlin—Göttingen—Heidelberg, 1958).

BURGESS, D. C. J. [1] Generalized intervals in partially ordered groups, *Proc. Cambridge Philos. Soc.*, **55** (1959), 165—171.

BUSULINI, B. [1] Sulla relazione triangolare in un *l*-gruppo, *Rend. Sem. Mat. Univ. Padova*, **28** (1958), 68—70.

CARTAN, H. [1] Un théorème sur les groupes ordonnés, *Bull. Sci. Math.*, **63** (1939), 201—205.

CERTAINE, J. [1] *Lattice-ordered groupoids and some related problems* (Harvard Doctoral Thesis, 1943).

CHEHATA, C. G. [1] An algebraically simple ordered group, *Proc. London Math. Soc.*, **2** (1952), 183—197. — [2] On an ordered semigroup, *Journ. London Math. Soc.*, **28** (1953), 353—356. — [3] On a theorem on ordered groups, *Proc. Glasgow Math. Ass.*, **4** (1958), 16—21.

CHOE, T. H. [1] The interval topology of a lattice-ordered group, *Kyungpook Math. Journ.*, **1** (1958), 69—74.

CLIFFORD, A. H. [1] Partially ordered Abelian groups, *Annals Math.*, **41** (1940), 465—473. — [2] A noncommutative ordinally simple linearly ordered group, *Proc. Amer. Math. Soc.*, **2** (1951), 902—903. — [3] A class of partially ordered Abelian groups related to Ky Fan's characterizing subgroups, *Amer. Journ. Math.*, **74** (1952), 347—356. — [4] Note on Hahn's theorem on ordered Abelian groups, *Proc. Amer. Math. Soc.*, **5** (1954), 860—863. — [5] Naturally totally ordered commutative semigroups, *Amer. Journ. Math.*, **76** (1954), 631—646. — [6] Ordered commutative semigroups of the second kind, *Proc. Amer. Math. Soc.*, **9** (1958), 682—687. — [7] Totally ordered commutative semigroups, *Bull. Amer. Math. Soc.*, **64** (1958), 305—316. — [8] Completion of semi-continuous ordered commutative semigroups, *Duke Math. Journ.*, **26** (1959), 41—59.

COHEN, L. W.—GOFFMAN, C. [1] The topology of ordered Abelian groups, *Trans. Amer. Math. Soc.*, **67** (1949), 310—319.

COHN, P. M. [1] Groups of order automorphisms of ordered sets, *Mathematika*, **4** (1957), 41—50.

CONRAD, P. [1] Embedding theorems for Abelian groups with valuations, *Amer. Journ. Math.*, **75** (1953), 1—29. — [2] On ordered division rings, *Proc. Amer. Math. Soc.*, **5** (1954), 323—328. — [3] Extensions of ordered groups, *Proc. Amer. Math. Soc.*, **6** (1955), 516—528. — [4] Methods of ordering a vector space, *Journ. Indian Math. Soc.*, **22** (1958), 1—25. — [5] On ordered vector spaces, *Journ. Indian Math. Soc.*, **22** (1958), 27—32. — [6] The group of order preserving automorphisms of an ordered Abelian group, *Proc. Amer. Math. Soc.*, **9** (1958), 382—389. — [7] A note on valued linear spaces, *Proc. Amer. Math. Soc.*, **9** (1958), 646—647. — [8] A correction and improvement of a theorem on ordered groups, *Proc. Amer. Math. Soc.*, **10** (1959), 182—184. — [9] Non-Abelian ordered groups, *Pacific Journ. Math.*, **9** (1959), 25—41. — [10] Right-ordered groups, *Michigan Math. Journ.*, **6** (1959), 267—275. — [11] Ordered semigroups, *Nagoya Math. Journ.*, **16** (1960), 51—64. — [12] The structure of a lattice-ordered group with a finite number of disjoint elements, *Michigan Math. Journ.*, **7** (1960), 171—180. — [13] Semigroups of real numbers, *Portugaliae Math.*, **18** (1959), 199—201. — [14] Some structure theorems for lattice-ordered groups, *Trans. Amer. Math. Soc.*, **99** (1961), 212—240.

CONRAD, P.—CLIFFORD, A. H. [1] Lattice-ordered groups having at most two disjoint elements, *Proc. Glasgow Math. Ass.*, **4** (1960), 111—113.

COTLAR, M.—ZARANTONELLO, E. [1] Semiordered groups and Riesz—Birkhoff *L*-ideals, *Fac. Ci. Mat. Univ. Nac. Lit.*, **8** (1948), 105—192. (Spanish.)

CRISTESCU, R. [1] La notion de composantes dans un groupe dirigé, *C. R. Acad. Sci. Paris*, **247** (1958), 1700—1702. — [2] Sur les groupes dirigés, *Czechosl. Math. Journ.*, **10** (1960), 17—26.

DIEUDONNÉ, J. [1] Sur la théorie de la divisibilité, *Bull. Soc. Math. France*, **69** (1941), 133—144. — [2] Sur les corps ordonnables, *Bol. Soc. Mat. São Paolo*, **1** (1946), 69—75; **2** (1947), 35.

DILWORTH, R. P. [1] Abstract residuation over lattices, *Bull. Amer. Math. Soc.*, **44** (1938), 262—268. — [2] Non-commutative residuated lattices, *Trans. Amer. Math. Soc.*, **46** (1939), 426—444.

DUBOIS, D. W. [1] On partly ordered fields, *Proc. Amer. Math. Soc.*, **7** (1956), 918—930.

DUBREIL, P. [1] *Algèbre*, I. (Paris, 1954). — [2] Introduction à la théorie des demi-groupes ordonnés, *Convegno Ital.-Franc. Algebra Astratta*

(Padova, 1956), 1—33. — [3] Quelques problèmes d'algèbre liés à la théorie des demi-groupes, *Colloque d'Algèbre Sup.* (Bruxelles, 1956), 29—44.

DUBREIL-JACOTIN, M.-L. [1] Quelques propriétés des équivalences régulières par rapport à la multiplication et à l'union, dans un treillis à multiplication commutative avec élément unitée, *C. R. Acad. Sci. Paris,* **232** (1951), 287—289. — [2] Quelques propriétés arithmétiques dans un demi-groupe demi-réticulé entier, *C. R. Acad. Sci. Paris,* **232** (1951), 1174—1176. — [3] Théorèmes de décomposition dans certains treillis et demi-groupes réticulés sans condition de chaîne, *C. R. Acad. Sci. Paris,* **234** (1952), 2415—2417.

DUBREIL-JACOTIN, M.-L.—CROISOT, R. [1] Équivalences régulières dans un ensemble ordonné, *Bull. Soc. Math. France,* **80** (1952), 11—35.

DUBREIL-JACOTIN, M.-L.—LESIEUR, L.—CROISOT, R. [1] *Leçons sur la théorie des treillis, des structures algébriques ordonnées et des treillis géométriques* (Paris, 1953), Partie II.

ERDŐS, J. [1] On the structure of ordered real vector spaces, *Publ. Math. Debrecen,* **4** (1956), 334—343.

EVERETT, C. J. [1] Sequence completion of lattice moduls, *Duke Math. Journ.,* **11** (1944), 109—119. — [2] Note on a result of L. Fuchs on ordered groups, *Amer. Journ. Math.,* **72** (1950), 216.

EVERETT, C. J.—ULAM, S. [1] On ordered groups, *Trans. Amer. Math. Soc.,* **57** (1945), 208—216.

FAN, K. [1] Partially ordered additive groups of continuous functions, *Annals Math.,* **51** (1950), 409—427.

FLEISCHER, I. [1] Remark on a theorem of Michiura, *Portugaliae Math.,* **12** (1953), 133. — [2] Functional representation of partially ordered groups, *Annals Math.,* **64** (1956), 260—263.

FOSTER, A. L. [1] Some elementary identities of ordered Abelian sets, *Amer. Math. Monthly,* **57** (1950), 681—683.

FREUDENTHAL, H. [1] Teilweise geordnete Moduln, *Proc. Nederl. Akad. Wetensch.,* **39** (1936), 641—651.

FUCHS, L. [1] Absolutes in partially ordered groups, *Proc. Nederl. Akad. Wetensch.,* **52** (1949), 251—255. — [2] On the extension of the partial order of groups, *Amer. Journ. Math.,* **72** (1950), 191—194. — [3] On partially ordered groups, *Proc. Nederl. Akad. Wetensch.,* **53** (1950), 828—834. — [4] The meet-decomposition of elements in lattice-ordered semi-groups, *Acta Sci. Math. Szeged,* **12 A** (1950), 105—111. — [5] The extension of partially ordered groups, *Acta Math. Acad. Sci. Hung.,* **1** (1950), 118—124. — [6] On mean systems, *Acta Math. Acad. Sci. Hung.,* **1** (1950), 303—320. — [7] The Zappa extension of partially ordered groups, *Proc. Nederl. Akad. Wetensch.,* **55** (1952), 363—368. — [8] A lattice-theoretic discussion of some problems in additive ideal theory, *Acta Math. Acad. Sci. Hung.,* **5** (1954), 299—313. — [9] Note on ordered groups and rings, *Fund. Math.,* **46** (1958), 167—174. — [10] Note on fully ordered semigroups, *Acta Math. Acad. Sci. Hung.,* **12** (1961), 255—259. — [11] On the ordering of quotient rings and quotient semigroups, *Acta Sci. Math. Szeged,* **22** (1961), 42—45.

GOFFMAN, C. [1] A lattice homomorphism of a lattice-ordered group, *Proc. Amer. Math. Soc.,* **8** (1957), 547—550. — [2] Remarks on lattice-ordered groups and vector lattices. I. Carathéodory functions, *Trans. Amer. Math. Soc.,* **88** (1958), 107—120. — [3] A class of lattice-ordered algebras, *Bull. Amer. Math. Soc.,* **64** (1958), 170—173.

GRAVETT, K. A. H. [1] Valued linear spaces, *Quart. Journ. Math. Oxford*, **6** (1955), 309—315. — [2] Ordered Abelian groups, *Quart. Journ. Math. Oxford*, **7** (1956), 57—63.

GRÄTZER, G.—SCHMIDT, E. T. [1] Über die Anordnung von Ringen, *Acta Math. Acad. Sci. Hung.*, **8** (1957), 259—260.

HAHN, H. [1] Über die nichtarchimedischen Grössensysteme, *S.-B. Akad. Wiss. Wien. IIa*, **116** (1907), 601—655.

HAUSNER, M.—WENDEL, J. G. [1] Ordered vector spaces, *Proc. Amer. Math. Soc.*, **3** (1952), 977—982.

HIGMAN, G. [1] Ordering by divisibility in abstract algebras, *Proc. London Math. Soc.*, **2** (1952), 326—336.

HILBERT, D. [1] *Grundlagen der Geometrie*, 7th ed. (Leipzig, 1930).

(HION) Х и о н, Я. В. [1] Архимедовски упорядоченные кольца (Archimedean ordered rings), У с п е х и М а т. Н а у к, **9** : 4 (1954), 237—242. — [2] Упорядоченные ассоциативные кольца (Ordered associative rings), Д о к л. А к а д. Н а у к, **101** (1955), 1005—1007. — [3] Упорядоченные полугруппы (Ordered semigroups), И з в. А к а д. Н а у к СССР, **21** (1957), 209—222.

HOFMANN, K. H. [1] Über archimedisch angeordnete, einseitig distributive Doppelloops, *Archiv Math.*, **10** (1959), 348—355.

HOLLAND, CH. [1] A totally ordered integral domain with a convex left ideal which is not an ideal, *Proc. Amer. Math. Soc.*, **11** (1960), 703.

HÖLDER, O. [1] Die Axiome der Quantität und die Lehre vom Mass, *Ber. Verh. Sächs. Ges. Wiss. Leipzig, Math. Phys. Cl.*, **53** (1901), 1—64.

HUNTINGTON, E. V. [1] A complete set of postulates for the theory of absolute continuous magnitude, *Trans. Amer. Math. Soc.*, **3** (1902), 264—279. — [2] Complete sets of postulates for the theories of positive integral and of positive rational numbers, *Trans. Amer. Math. Soc.*, **3** (1902), 280—284.

ISÉKI, K. [1] Structure of special ordered loops, *Portugaliae Math.*, **10** (1951), 81—83. — [2] On simply ordered groups, *Portugaliae Math.*, **10** (1951), 85—88.

ISIWATA, T. [1] Non-discrete linearly ordered groups, *Kōdai Math. Sem. Reports*, **1950**, 84—88. — [2] Linearization of topological groups and ordered rings, *Kōdai Math. Sem. Reports*, **1952**, 33—35.

IWASAWA, K. [1] On the structure of conditionally complete lattice-groups, *Japan. Journ. Math.*, **18** (1943), 777—789. — [2] On linearly ordered groups, *Journ. Math. Soc. Japan*, **1** (1948), 1—9.

JAEGER, A. [1] Adjunction of subfield closures to ordered division rings, *Trans. Amer. Math. Soc.*, **73** (1952), 35—39.

JAFFARD, P. [1] Théorie des filets dans les groupes réticulés, *C. R. Acad. Sci. Paris*, **230** (1950), 1024—1025. — [2] Applications de la théorie des filets, *C. R. Acad. Sci. Paris*, **230** (1950), 1125—1126. — [3] Nouvelles applications de la théorie des filets, *C. R. Acad. Sci. Paris*, **230** (1950), 1631—1632. — [4] Groupes archimédiens et para-archimédiens, *C. R. Acad. Sci. Paris*, **231** (1950), 1278—1280. — [5] Théorie arithmétique des anneaux du type de Dedekind. I—II, *Bull. Soc. Math. France*, **80** (1952), 61—100; **81** (1953), 41—61. — [6] Contribution à l'étude des groupes ordonnés, *Journ. Math. Pures Appl.*, **32** (1953), 203—280. — [7] Extension des groupes réticulés et applications, *Publ. Sci. Univ. Alger.*, Sér. A, **1** (1954), 197—222. — [8] Réalisation des groupes complètement réticulés, *Bull. Soc. Math. France*, **84** (1956), 295—305. — [9] Sur les groupes réticulés associés à un groupe ordonné, *Publ. Sci. Univ. Alger.*,

Sér. A, **2** (1957), 173—203. — [10] *Les systèmes d'idéaux* (Paris, 1960).

JAKUBÍK, J. [1] О главных идеалах в структурно упорядоченных группах (Principal ideals in *l*-groups), *Czechosl. Math. Journ.,* **9** (1959), 528—543. — [2] Konvexe Ketten in *l*-Gruppen, *Časopis Pěst. Mat.,* **84** (1959), 53—63. — [3] Об одном классе структурно упорядоченных групп (On a class of *l*-groups), *Časopis Pěst. Mat.,* **84** (1959), 150—161. — [4] Konvexe Ketten in halbgeordneten Gruppen, *Matematicko-Fyzikálny Časopis Slov. Akad. Vied,* **9** (1959), 236—242. (Slovak.) — [5] Об одном свойстве структурно упорядоченных групп (Über eine Eigenschaft von *l*-Gruppen), *Časopis Pěst. Mat.,* **85** (1960), 51—59. — [6] Прямые разложения частично упорядоченных групп (Direkte Zerlegungen der teilweise geordneten Gruppen), *Czechosl. Math. Journ.,* **10** (1960), 231—243.

JOHNSON, D. G. [1] A structure theory for a class of lattice-ordered rings, *Acta Math.,* **104** (1960), 163—215.

JOHNSON, R. E. [1] On ordered domains of integrity, *Proc. Amer. Math. Soc.,* **3** (1952), 414—416.

KALMAN, J. A. [1] An identity for *l*-groups, *Proc. Amer. Math. Soc.,* **7** (1956), 931—932.

KANTOROVITCH, L. V. [1] Lineare halbgeordnete Räume, М а т. С б о р-н и к, **2** (44) (1937), 121—168.

KERSTAN, J. [1] Elementfreie Begründung der allgemeinen Ideal- und Modultheorie, *Ber. Math. Tagung Berlin,* **1953**, 49—57.

KLEIN-BARMEN, F. [1] Über Verbände mit einer weiteren assoziativen und kommutativen Elementverknüpfung, *Math. Zschrift,* **47** (1942), 85—104. — [2] Über gewisse Halbverbände und kommutative Semigruppen. I—II, *Math. Zschrift,* **48** (1942—43), 275—288, 715—734. — [3] Ein Beitrag zur Theorie der linearen Holoide, *Math. Zschrift,* **51** (1949), 355—366.

KLINGENBERG, W. [1] Sopra il numero degli ordinamenti di un corpo, *Atti Accad. Naz. Lincei,* **14** (1953), 395—396.

(KONTOROVIČ—KUTYEV) К о н т о р о в и ч, П. Г. — К у т ы е в, К. М. [1] К теории структурно упорядоченных групп (On the theory of lattice-ordered groups), И з в. В ы с ш. У ч е б н ы х З а в е д. М а т е м., **1959**, No. 3, 112—120.

KRISHNAN, V. S. [1] L'extension d'une (<,.) algèbre à une (Σ^*,.) algèbre. I—II, *C. R. Acad. Sci. Paris,* **230** (1950), 1447—1448, 1559—1561. — [2] Les algèbres partiellement ordonnées et leurs extensions, *Bull. Soc. Math. France,* **78** (1950), 235—263; **79** (1951), 85—120.

KRULL, W. [1] Allgemeine Bewertungstheorie, *Journ. reine u. angew. Math.,* **167** (1932), 160—196. — [2] Halbgeordnete Gruppen und asymptotische Grössenordnung, *Archiv Math.,* **3** (1952), 1—7. — [3] Über geordnete Gruppen von reellen Funktionen, *Math. Zschrift,* **64** (1955), 10—40. — [4] Über die Endomorphismen von total geordneten archimedischen abelschen Gruppen, *Math. Zschrift,* **74** (1960), 81—90.

KUDLAČEK, V. [1] Lattice-ordered groupoids, *Časopis Pěst. Mat.,* **80** (1955), 44—50.

(KUTYEV) К у т ы е в, К. М. [1] О регулярных структурно упорядоченных группах (On regular lattice-ordered groups), У с п е х и М а т. Н а у к, **11** : 1 (1956), 256. — [2] К теории частично упорядоченных групп (On the theory of partially ordered groups), У с п е х и М а т. Н а у к, **11** : 1 (1956), 258. — [3] ПС-изоморфизмы частично упоря-

доченных локально нильпотентных групп (SL-isomorphisms of partially oideied locally nilpotent groups), Успехи Мат. Наук,
11 : 2 (1956), 193—198. — [4] К теории структурно упорядоченных
групп (On the theory of lattice-ordered groups), Успехи Мат.
Наук, **13** : 3 (1958), 238—239.

LESIEUR, L. [1] Sur les treillis multiplicatifs complets à condition
minimale, *C. R. Acad. Sci. Paris*, **232** (1951), 290—292. — [2]
Conditions suffisantes pour que, dans un treillis multiplicatif com-
plet, la condition de chaîne descendante entraîne la condition de
chaîne ascendante, *C. R. Acad. Sci. Paris*, **234** (1952), 1017—1019.
— [3] Théorèmes de décomposition dans certains demi-groupes
réticul s satisfaisant à la condition de chaîne descendante affaiblie,
C. R. Acad. Sci. Paris, **234** (1952), 2250—2252. — [4] Sur les demi-
groupes réticulés satisfaisant à une condition de chaîne, *Bull. Soc.
Math. France*, **83** (1955), 161—193.

LESIEUR, L.—CROISOT, R. [1] Théorie Noethérienne des anneaux, des
demi-groupes et des modules dans le cas non commutatif. I: *Colloque d'Algèbre Supérieure* (Bruxelles, 1956), 79—121; II: *Math.
Annalen*, **134** (1958), 458—476; III: *Bull. Acad. Royale Belgique*,
44 (1958), 75—93.

LEVI, F. W. [1] Arithmetische Gesetze im Gebiete diskreter Gruppen,
Rend. Palermo, **35** (1913), 225—236. — [2] Ordered groups, *Proc.
Indian Acad. Sci.*, *Sect. A*, **16** (1942), 256—263. — [3] Contributions
to the theory of ordered groups, *Proc. Indian Acad. Sci.*, *Sect. A*,
17 (1943), 199—201.

LINÉS ESCARDÓ, E.—MALLOL BALMAÑA, R. [1] On *l*-groups, *Revista
Mat. Hisp.-Amer.*, **12** (1952), 129—136, 137.

LOONSTRA, F. [1] Ordered groups, *Proc. Nederl. Akad. Wetensch.*, **49**
(1946), 41—46. — [2] The classes of ordered groups, *Proc. Internat.
Congr. Math.* (Cambridge, 1950), vol. **1**. — [3] The classes of partially
ordered groups, *Compositio Math.*, **9** (1951), 130—140. — [4]
Discrete groups, *Proc. Nederl. Akad. Wetensch.*, **54** (1951), 162—
168. — [5] L'extension du groupe ordonné des entiers rationnels
par le même groupe, *Proc. Nederl. Akad. Wetensch.*, **58** (1955),
41—49.

LORENZ, K. [1] Über Strukturverbände von Verbandsgruppen, *Acta
Math. Acad. Sci. Hung.*, **13** (1962), 55—67.

LORENZEN, P. [1] Abstrakte Begründung der multiplikativen Ideal-
theorie, *Math. Zschrift*, **45** (1939), 533—553. — [2] Über halbgeordnete Gruppen, *Archiv Math.*, **2** (1949), 66—70. — [3] Über
halbgeordnete Gruppen, *Math. Zschrift*, **52** (1950), 483—526. —
[4] Die Erweiterung halbgeordneter Gruppen zu Verbandsgruppen,
Math. Zschrift, **58** (1953), 15—24.

ŁOŚ, J. [1] On the existence of linear order in a group, *Bull. Acad.
Polon. Sci. Cl. III*, **2** (1954), 21—23.

(LYAPIN) Ляпин, Е. С. [1] Полугруппы (*Semigroups*) (Москва,
1961).

(MAL'CEV) Мальцев, А. И. [1] О включении групповых алгебр
в алгебры с делением (On the embedding of group algebras in
division algebras), Докл. Акад. Наук СССР, **60** (1948),
1499—1501. — [2] Об упорядоченных группах (On ordered groups),
Изв. Акад. Наук СССР, **13** (1949), 473—482. — [3] О
доупорядочении групп (On the ordering of groups), Труды
Мат. Инст. Стеклова, **38** (1951), 173—175. — [4]
Замечание о частично упорядоченных группах (Remark on par-

tially ordered groups), Учен. Зап. Пед. Инст. Иваново, 10 (1956), 3—5.

MATSUSHITA, S. [1] On the foundation of orders in groups, *Journ. Inst. Polytechn., Osaka City Univ.*, 2 (1951), 19—22. — [2] Sur la puissance des ordres dans un groupe libre, *Proc. Nederl. Akad. Wetensch.*, 56 (1953), 15—16.

MICHIURA, T. [1] On a definition of lattice-ordered groups. I—II. *Journ. Osaka Inst. Sci. Techn.*, 1 (1949), 27, 117—119. — [2] Lattice-ordered rings and ordered characterizations of integers, *Journ. Osaka Inst. Sci. Techn.*, 1 (1949), 29—31. — [3] Sur les groupes semi-ordonnés, *C. R. Acad. Sci. Paris*, 231 (1950), 1403—1404. — [4] On simply ordered groups, *Portugaliae Math.*, 10 (1951), 89—95. — [5] Remark on a representation of simply ordered groups, *Proc. Nederl. Akad. Wetensch.*, 54 (1951), 386—387. — [6] Commutativity in simply ordered groups, *Journ. Osaka Inst. Sci. Techn.*, 3 (1951), 39—41. — [7] Sur les groupes ordonnés. II—III, *C. R. Acad. Sci. Paris*, 234 (1952), 1422—1423, 1521—1522. — [8] On partially ordered groups without proper convex subgroups, *Proc. Nederl. Akad. Wetensch.*, 56 (1952), 231—232.

MITRINOVITS, D. S. [1] Sur certaines relations restant valables si l'on permute les operateurs y intervenant, *Bull. Soc. Math. Phys. Serbie*, 8 (1956), 15—22.

MOLINARO, I. [1] Généralisation de l'équivalence d'Artin, *C. R. Acad. Sci. Paris*, 238 (1954), 1284—1286, 1767—1769. — [2] Demi-groupes résidutifs, *Thèse* (Paris, 1956).

MONNA, A. F. [1] On ordered groups and linear spaces, *Nederl. Akad. Wetensch. Verslagen Afd. Natuurkunde*, 63 (1944), 178—182. (Dutch.)

MOUFANG, R. [1] Einige Untersuchungen über geordnete Schiefkörper, *Journ. reine u. angew. Math.*, 176 (1937), 203—223.

MURATA, K. [1] A theorem on residuated lattices, *Proc. Japan Acad.*, 33 (1957), 639—641. — [2] Decomposition of radical elements of a commutative residuated lattice, *Journ. Inst. Polytechnics, Osaka Univ.*, 10 (1959), 31—34. — [3] Additive ideal theory in multiplicative systems, *Journ. Inst. Polytechnics, Osaka Univ.*, 10 (1959), 91—115.

NAKADA, O. [1] Partially ordered Abelian semigroups. I—II, *Journ. Fac. Sci. Hokkaido Univ.*, 11 (1951), 181—189; 12 (1952), 73—86.

NAKAMURA, M. [1] Partially ordered rings, *Tohoku Math. Journ.*, 47 (1940), 251—254.

NAKANO, H. [1] Teilweise geordnete Algebra, *Proc. Imp. Acad. Tokyo*, 16 (1940), 437—441, and *Japan. Journ. Math.*, 17 (1941), 425—511.

NAKAYAMA, K. [1] Note on lattice-ordered groups, *Proc. Imp. Acad. Tokyo*, 18 (1942), 1—4. — [2] On Krull's conjecture concerning completely integrally closed integrity domains. I. *Proc. Imp. Acad. Tokyo*, 18 (1942), 185—187.

NEUMANN, B. H. [1] On ordered groups, *Amer. Journ. Math.*, 71 (1949), 1—18. — [2] On ordered division rings, *Trans. Amer. Math. Soc.*, 66 (1949), 202—252. — [3] An embedding theorem for algebraic systems, *Proc. London Math. Soc.*, 4 (1954), 138—153. — [4] Embedding theorems for ordered groups, *Journ. London Math. Soc.*, 35 (1960), 503—512.

NEUMANN, B. H.—SHEPPERD, J. A. H. [1] Finite extensions of fully ordered groups, *Proc. Royal Soc. London, Ser. A*, 239 (1957), 320—327.

OGASAWARA, T. [1] Commutativity of Archimedean semiordered groups, *Journ. Sci. Hiroshima Univ., Ser. A*, 12 (1948), 249—254. (Japanese.)

Ohnishi, M. [1] On linearization of ordered groups, *Osaka Math. Journ.*, **2** (1950), 161—164. — [2] Linear-order on a group, *Osaka Math. Journ.*, **4** (1952), 17—18.

Pickert, G. [1] *Einführung in die höhere Algebra* (Göttingen, 1951). — [2] *Projektive Ebenen* (Berlin—Göttingen—Heidelberg, 1955).

Pierce, R. S. [1] Homomorphisms of semi-groups, *Annals Math.*, **59** (1954), 287—291. — [2] Radicals in function rings, *Duke Math. Journ.*, **23** (1956), 253—261.

(Pinsker) Пинскер, А. Г. [1] Разложение полуупорядоченных групп и пространств (Decomposition of semiordered groups and spaces), Учен. Зап. Пед. Инст. Ленинград, **86** (1949), 235—284. — [2] Расширение полуупорядоченных групп и пространств (Extension of semiordered groups and spaces), Учен. Зап. Пед. Инст. Ленинград, **86** (1949), 285—315.

(Podderyugin) Поддерюгин, В. Д. [1] Условие упорядочиваемости произвольного кольца (A condition of orderability for an arbitrary ring), Успехи Мат. Наук, **9**:4 (1954), 211—216. — [2] Условия упорядочиваемости группы (A condition of orderability for a group), Изв. Акад. Наук СССР, **21** (1957), 199—208.

Prenowitz, W. [1] Partially ordered fields and geometries, *Amer. Math. Monthly*, **53** (1946), 439—449.

Rédei, L. [1] *Algebra*. I (Budapest, 1954 or Leipzig, 1959).

Ree, R. [1] On ordered, finitely generated, solvable groups, *Trans. Royal Soc. Canada*, (3), **48** (1954), 39—42.

Ribenboim, P. [1] Conjonction d'ordres dans les groupes abéliens ordonnés, *Anais Acad. Brasil. Ci.*, **29** (1957), 201—224. — [2] Sur les groupes totalement ordonnés et l'arithmétique des anneaux de valuation, *Summa Brasil. Math.*, **4** (1958), 1—64. — [3] Sur quelques constructions de groupes réticulés et l'équivalence logique entre l'affinement de filtres et d'ordres, *Summa Brasil. Math.*, **4** (1958), 65—89. — [4] Un théorème de réalisation de groupes réticulés, *Pacific Journ. Math.*, **10** (1960), 305—308.

Rieger, L. S. [1] On the ordered and cyclically ordered groups. I—III, *Věstník Král. České Spol. Nauk*, **1946**, no. 6, 1—31; **1947**, no. 1, 1—33; **1948**, no. 1, 1—26. (Czech.)

Riesz, F. [1] Sur quelques notions fondamentales dans la théorie générale des opérations linéaires, *Annals Math.*, **41** (1940), 174—206.

Robinson, A. [1] Note on an embedding theorem for algebraic systems, *Journ. London Math. Soc.*, **30** (1955), 249—252. — [2] On ordered fields and definite functions, *Math. Ann.*, **130** (1955), 257—271.— [3] Further remarks on ordered fields and definite functions, *Math. Ann.*, **130** (1956), 405—409.

Schilling, O. F. G. [1] *The theory of valuations* (New York, 1950).

Serre, J.-P. [1] Extensions de corps ordonnés, *C. R. Acad. Sci. Paris*, **229** (1949), 576—577.

Shepperd, J. A. H. [1] Transitivities of betweenness and separation and the definitions of betweenness and separation groups, *Journ. London Math. Soc.*, **31** (1956), 240—248. — [2] Betweenness groups, *Journ. London Math. Soc.*, **32** (1957), 277—285. — [3] Separation groups, *Proc. London Math. Soc.*, **7** (1957), 518—548.

Šik, F. [1] К теории структурно упорядоченных групп (Zur Theorie der halbgeordneten Gruppen), *Czechosl. Math. Journ.*, **6** (1956), 1—25.— [2] Über Summen einfach geordneter Gruppen, *Czechosl. Math. Journ.*, **8** (1958), 22—53. — [3] Automorphismen geordneter

Mengen, *Časopis Pěst. Mat.*, **83** (1958), 1—22. — [4] Über subdirekte Summen geordneter Gruppen, *Czechosl. Math. Journ.*, **10** (1960), 400—424. — [5] Erweiterungen teilweise geordneter Gruppen, *Publ. Fac. Sci. Univ. Brno*, no. 410 (1960), 65—80.

Sikorski, R. [1] On an ordered algebraic field, *C. R. Soc. Sci. Lettres de Varsovie*, Cl. III, **41** (1948), 69—96. — [2] On algebraic extensions of ordered fields, *Ann. Soc. Polonaise Math.*, **22** (1950), 173—184.

(Šimbireva) Шимбирева, Е. П. [1] К теории частично упорядоченных групп (On the theory of partially ordered groups), Мат. Сборник, **20** (1947), 145—178.

(Smirnov) Смирнов, Д. М. [1] Инфраинвариантные подгруппы (Infrainvariant subgroups), Учен. Зап. Пед. Инст. Иваново, **4** (1953), 9z—96.

(Sorokina) Сорокина, В. И. [1] Понятие группы и линейного множества с дизъюнктными элементами (The concept of a group and a linear set with disjoint elements), Учен. Зап. Пед. Инст. Ленинград, **103** (1955), 179—207.

Steinfeld, O. [1] Verbandstheoretische Betrachtung gewisser idealtheoretischer Fragen, *Acta Sci. Math. Szeged*, **22** (1961), 136—149.

(Stelleckiǐ) Стеллецкий, И. В. [1] Нильпотентные структуры (Nilpotent lattices), Труды Моск. Мат. Общ., **9** (1960), 211—235.

Stone, M. H. [1] A general theory of spectra. I—II, *Proc. Nat. Acad. Sci. USA*, **26** (1940), 280—283; **27** (1941), 83—87. — [2] Pseudonorms and partial orderings in Abelian groups, *Annals Math.*, **48** (1949), 851—856.

(Sul′geǐfer) Шульгейфер, Е. Г. [1] Разложение на простые множители в структурах с умножением (Decomposition into prime factors in lattices with multiplication), Укр. Мат. Журн., **2** : 3 (1950), 100—114.

Świerczkowski, S. [1] On cyclically ordered groups, *Fund. Math.*, **47** (1959), 161—166. — [2] On cyclically ordered intervals of integers, *Fund. Math.*, **47** (1959), 167—172.

Szele, T. [1] On ordered skew fields, *Proc. Amer. Math. Soc.*, **3** (1952), 410—413.

Tallini, G. [1] Sui sistemi a doppia composizione ordinati archimedei, *Atti Accad. Naz. Lincei*, **18** (1955), 367—373.

Tamari, D. [1] Groupoïdes reliés et demi-groupes ordonnés, *C. R. Acad. Sci. Paris*, **228** (1949), 1184—1186. — [2] Groupoïdes ordonnés. L'ordre lexicographique pondéré, *C. R. Acad. Sci. Paris*, **228** (1949), 1909—1911. — [3] Ordres pondérés. Caractérisation de l'ordre naturel comme l'ordre du semi-groupe multiplicatif des nombres naturels, *C. R. Acad. Sci. Paris*, **229** (1949), 98—100. — [4] Monoïdes préordonnés et chaînes de Malcev, *Bull. Soc. Math. France*, **82** (1954), 53—96.

Tamura, T. [1] Commutative nonpotent Archimedean semigroups with cancellation law. I, *Journ. Gakugei Tokushima Univ.*, **8** (1957), 5—11.

Tarski, A. [1] Sur les groupes d'Abel ordonnés, *Annales Soc. Pol. Math.*, **7** (1929), 267—268. — [2] *Cardinal algebras* (New York, 1949).

Teh, H.-H. [1] Construction of orders in Abelian groups, *Proc. Cambridge Philos. Soc.*, **57** (1961), 476—482.

(Terehov) Терехов, А. А. [1] О вполне доупорядочиваемых группах (On completely orderable groups), Докл. Акад. Наук СССР, **129** (1959), 34—36.

TREVISAN, G. [1] Sulla equivalenza archimedea relative alle gruppo-strutture, *Rend. Sem. Mat. Univ. Padova*, **20** (1951), 425—429. — [2] Classificazione dei semplici ordinamenti di un gruppo libero commutativo con *n* generatori, *Rend. Sem. Mat. Univ. Padova*, **22** (1953), 143—156.

(VAGNER) Вагнер, В. В. [1] Представление упорядоченных полугрупп (Representations of ordered semigroups), Мат. Сборник, **38** (1956), 203—240.

(VINOGRADOV) Виноградов, А. А. [1] О свободном произведении упорядоченных групп (On the free product of ordered groups), Мат. Сборник, **25** (1949), 163—168. — [2] Частично упорядоченные локально нильпотентные группы (Partially ordered locally nilpotent groups), Учен. Зап. Пед. Инст. Иваново, **4** (1953), 3—18. —[3] К теории упорядоченных полугрупп (On the theory of ordered semigroups), Учен. Зап. Пед. Инст. Иваново, **4** (1953), 19—21. — [4] К теории частично упорядоченных нильпотентных групп (On the theory of partially ordered nilpotent groups), Учен. Зап. Пед. Инст. Иваново, **5** (1954), 61—64.

VAN DER WAERDEN, B. L. [1] *Moderne Algebra*, vol. I (Berlin—Göttingen—Heidelberg, 1950).

WAGNER, W. [1] Über die Grundlagen der projektiven Geometrie und allgemeine Zahlensysteme, *Math. Ann.*, **113** (1937), 528—567.

WANG, SH.-CH. [1] Representation of ordered Abelian groups and ordered rings of finite degree, *Acta Math. Sinica*, **5** (1955), 425—432. — [2] A note on ordered rings of real vectors, *Acta Math. Sinica*, **5** (1955), 65—80. (Chinese.)

WARD, W. [1] Residuated distributive lattices, *Duke Math. Journ.*, **6** (1940), 641—651.

WARD, W.—DILWORTH, R. P. [1] Residuated lattices, *Proc. Nat. Acad. Sci. USA*, **24** (1938), 162—164. — [2] Residuated lattices, *Trans. Amer. Math. Soc.*, **45** (1939), 335—354.

YAMADA, M. [1] Regularly totally ordered semigroups. I, *Science Reports of Shimane Univ.*, **1957**, 14—23.

YOSIDA, K. [1] On vector lattice with a unit. *Proc. Imp. Acad. Tokyo*, **17** (1941), 121—124.

YOSIDA, K.—FUKAMIYA, M. [1] On vector lattice with a unit. II, *Proc. Imp. Acad. Tokyo*, **17** (1941), 479—482.

(ZAĬCEVA) Зайцева, М. И. [1] О совокупности упорядочений абелевой группы (On the set of orderings of Abelian groups), Усп. Мат. Наук, **8** : 1 (1953), 135—137. — [2] Правоупорядоченные группы (Right-ordered groups), Учен. Зап. Шуйск. Пед. Инст., **6** (1958), 205—226.

ZAMANSKY, M. [1] Groupes de Riesz, *C. R. Acad. Sci. Paris*, **248** (1959), 2933—2934.

(ZARECKĬ) Зарецкий, К. А. [1] Представление упорядоченных полугрупп бинарными отношениями (Representation of ordered semigroups by binary relations), Изв. Акад. Наук СССР, **23** (1959), 48—50.

ZELINSKY, D. [1] On ordered loops, *Amer. Journ. Math.*, **70** (1948), 681—697. — [2] Non-associative valuations, *Bull. Amer. Math. Soc.*, **54** (1948), 175—183.

ZEMMER, J. L. [1] Ordered algebras which contain divisors of zero, *Duke Math. Journ.*, **20** (1953), 177—183.

AUTHOR INDEX

SUBJECT INDEX